What the Digital Future Holds

The Digital Future of Management

Paul Michelman, series editor

How to Go Digital: Practical Wisdom to Help Drive Your Organization's Digital Transformation

What the Digital Future Holds: 20 Groundbreaking Essays on How Technology Is Reshaping the Practice of Management

.

What the Digital Future Holds

20 Groundbreaking Essays on How Technology Is Reshaping the Practice of Management

MIT Sloan Management Review

MITSloan
Management Review

The MIT Press
Cambridge, Massachusetts
London, England

This book was set in Neue Haas Grotesk and Stone Serif by the MIT Press. Printed and bound in the United States of America.

Library of Congress Cataloging-in-Publication Data is available.

ISBN: 978-0-262-53499-4

10 9 8 7 6 5 4 3 2 1

Contents

Series Foreword

Books in the Digital Future of Management series draw from the print and web pages of *MIT Sloan Management Review* to deliver expert insights and sharply tuned advice on navigating the unprecedented challenges of the digital world. These books are essential reading for executives from the world's leading source of ideas on how technology is transforming the practice of management.

Paul Michelman
Editor in chief
MIT Sloan Management Review

Introduction: Tales from the Digital Frontier

Paul Michelman

The impact of digital technology on how businesses design and produce goods, interact with their supply chains, manage internal communication, and connect with customers is a rich topic that has been and continues to be broadly addressed in both commercial and academic business media.

But as the digital revolution enters its next phase, we find ourselves confronting a new set of questions about the relationship between technology and management. These questions go to the core of the organization:

- How will big data inform hiring decisions?
- What happens to marketing when marketers can map consumers' brain patterns?
- Are new communication technologies really delivering on the promise to empower frontline workers, or are they unleashing organizational chaos?
- What role will algorithms play in creating corporate strategy?
- How do you give performance feedback to a machine? How will our robot managers provide it to us?

And as the influence of those who write the lines of software that helps us manage our lives and work continues to increase, who will watch the coders?

Technology pervades organizations across the globe, yet the organizational form has, thus far, transformed relatively little as a result; the way we lead organizations even less so. I believe that is about to change. Today, there is a growing wellspring of research and insight exploring technology's foundational impacts on management, emerging from both academia and industry.

Much of the conversation focuses on the evolving relationship between humans and cognitive technologies. Do continual advances in intelligent algorithms and learning machines represent an existential threat to management as we know it? Perhaps that overstates things a bit, but the tone of the conversation about technology's role in the organization is changing dramatically. We now speak of software as a colleague and a coworker. Collaboration between humans and machines no longer appears to be the stuff of science fiction.

As the contributors to this volume note, with great advancements come great responsibilities. We must take care in managing the effects of technological innovation on humans and humanity. The digital revolution will continue to challenge us with new forms of stress and with ethical and economic uncertainties. At the same time, we cannot shrink from our responsibility to harness technology's profound potential to improve our world.

Contributors to this volume represent a broad range of academic and professional disciplines. They are luminaries in the fields of corporate strategy, leadership, software development, entrepreneurship, engineering, operations, economics, talent development, sustainability, and more. There are few topics where such a broad range of experts can naturally converge: Digitization makes for strange bedfellows.

It is valuable—even necessary—to draw on a diversity of perspectives to understand how our digital world is evolving. Today, HR executives have more to learn from data experts than from talent experts. Marketing leaders should be studying machine learning. And CTOs need to focus on the human impact of their strategy decisions as never before.

We are fast approaching the corporate singularity, where the once discrete domains of technology and business are forever converged. But how prepared are we for such a time?

This book aims to make a small contribution to your state of readiness. The purpose of these essays—taken separately or together—is to help you see what is possible, probable, and developing with much greater speed than many of us could have ever imagined. It is your guide through the next wave of smart technology and the disruption it promises to bring to our organizations.

This book will help you see around the digital bend at a time when too many of us are chasing our analog tails.

1

Managing the Bots That Are Managing the Business

Tim O'Reilly

Science fiction writer William Gibson once said, "The future is already here. It's just not very evenly distributed." You don't need to wait five years to see how technology will change the practice of management. You just need to study companies that are already living in the future that remains around the corner for everyone else.

You must also reframe what you see so that you aren't blinded by what you already know. Consider Google, Facebook, Amazon.com, or a host of more recent Silicon Valley startups. They employ tens of thousands of workers. If you think with a 20th century factory mindset, those workers spend their days grinding out products, just like their industrial forebears; only today, they are producing software rather than physical goods. If, instead, you step back and view these organizations with a 21st century mindset, you realize that a large part of the work of these companies—delivering search results, news and information, social network status updates, and relevant products for purchase—is performed by software programs and algorithms. These programs are the workers, and the human software developers who create them are their managers.

Each day, these "managers" take in feedback about their electronic workers' performance—as measured in real-time data from the marketplace—and they provide feedback to the workers in the form of minor tweaks and updates to their programs or algorithms. The human managers also have their own managers, but hierarchies are often flat, and multiple levels of management are aligned around a set of data-driven objectives and key results (OKRs) that are measurable in a way that allows even the electronic "workers" to be guided by these objectives.

And if your notion of management isn't stretched enough, consider that the algorithms themselves are sometimes the managers—of human workers. Look at Uber Technologies Inc. and Lyft Inc. as compared to an old-school taxi service. Cars are dispatched not by a person but by a management program that collects passenger requests in real time and serves them to the closest available drivers. Humans are still in the loop both as workers delivering the actual service and as managers who set the rules of the algorithms, test whether they are working correctly, and adjust them when they are not. Human managers are called in to deal directly with the human workers, the drivers, only under rare circumstances.

While it is early in this process, you can make the case that the core function of management has gone from managing the business to managing the bots that are managing the business.

Some further implications of this insight include:

- A typical programmer in a 20th century IT shop was a worker building to a specification, not that different from a shop floor worker assembling a predefined product. A 21st century software developer is deeply engaged in product design and iterative, customer-focused development. For larger programs, this is a team exercise, and leadership means organizing a shared creative vision. Technology is not a back-office function. It is central to the management capability of the entire organization. And companies whose CEOs are also the chief product designers (think Larry Page of Alphabet Inc., Jeff Bezos of Amazon, or Apple under Steve Jobs) can outperform those whose leaders lack the capability to lead not just their human workers but their electronic workers as well.

- In 20th century organizations, you gained influence by gaining budget to hire more workers. In 21st century organizations, you gain effectiveness through your ability to *create* more workers—of the 21st century variety. Even in jobs that are not considered "programming jobs," the ability to create and marshal electronic resources is key to advancement. A salesperson who can write a bot to "scrape" LinkedIn for leads has an edge over someone who has to do it manually. A marketer who can build an online survey or data-gathering app has an advantage over one who has to hire an outside company. A designer/developer who can build a working application prototype is more valuable than a designer who is only able to draw a picture.

- Managers must become product and experience designers, deeply engaged with customers and their needs, creating services that start out as a compelling promise and get better over time the more people use them, via a "build-measure-learn" process. A service like Uber is based on a deep rethinking of the fundamental workflows of on-demand transportation (what used to be called "taxis") in light of what technology now makes possible. Before Uber, who would have thought that a passenger could summon a car to a specific spot and know just when they were going to be picked up? Yet that capability was already lying latent in smartphones.

- There is an arc to knowledge in which expertise becomes embodied in products. Workers can be "upskilled" not just by training but by software assistants that allow them to do jobs for which they were previously under-qualified. "The Knowledge," the legendary test that London taxicab drivers must pass, requires years of study, yet with the aid of Waze, Google Maps, and the Uber or Lyft app, virtually anyone can become a driver for hire, even in a strange city. There is a lot of fear about technology replacing workers, but in what I call "Next Economy companies," technology is used to create new opportunities by augmenting workers.

We are just at the beginning of the transformation from an economy dominated by human workers to one dominated by electronic workers. The great management challenge of the next few decades will be understanding how to get the best out of both humans and machines, and understanding the ins and outs of who manages whom.

2

Digital Today, Cognitive Tomorrow

Ginni Rometty

In today's economy, we are seeing companies, business models, products, and processes undergoing major transformation. Enterprises and governments are rapidly "becoming digital" as they seek to capture the cost savings, agility, and collaboration enabled by cloud, analytics, mobile, and social technologies.

However, digital is not the destination. Rather, it is laying the foundation for a much more profound transformation to come. Within five years, I believe all major business decisions will be enhanced by cognitive technologies.

I sensed the magnitude of the transition for the first time in 2011, when I watched IBM's Watson system win on "Jeopardy!" At the time, I felt that I was watching history in the making: The technology known as artificial intelligence (AI) was finally moving from the lab into the world.

Why are we seeing this now?

First, the technologies required for cognitive systems—not just AI, but a broad spectrum of capabilities that include natural language processing, human–computer interaction, deep learning, neural nets, and more—have made exponential advances in recent years.

Second, the abundance of data being generated throughout the world today requires cognitive technology. Much of this data is "unstructured": video, audio, sensor outputs, and everything we encode in language, from medical journals to tweets. However, such unstructured data are "dark" to traditional computer systems. Computers can capture, move, and store the data, but they cannot understand what the data mean (which is why cognitive systems are so vital).

Finally, and most important, we will see systems that learn. We *need* systems that learn. Think of the challenges and issues we face today: predicting risk in financial markets, anticipating consumer behavior, ensuring public safety, managing traffic, optimizing global supply chains, personalizing medicine, treating chronic diseases, and preventing pandemics.

The challenges today go beyond information overload. In many ways, we live in an era of cognitive overload, characterized by an exponential increase in the complexity of decision making. It's impossible to create protocols, algorithms, or software code to successfully anticipate all the potential permutations, trajectories, and interactions. But cognitive systems are not simply programmed. They actually improve with use, as they receive expert training, interact with clients and customers, and ingest data from their own experiences, successes, and failures.

Some people think of cognitive systems as supercomputers, and there is no question that the computational power behind systems like Watson is considerable. But thanks to the increasing prevalence of application program interfaces (APIs)—which can be encoded into digital services and easily accessed or combined in new ways in the cloud—it's possible to build a kind of thinking into virtually every digital application, product, and system.

And because we can, we will. If it's digital today, it will be cognitive tomorrow—and not a distant tomorrow. IDC Research Inc. has estimated that by 2018, more than half of the teams developing apps will embed some kind of cognitive services in them, up from 1% in 2015.

Cognitive systems are already transforming everything from the world-changing to the everyday. For example, cognitive oncology is a reality thanks to technology developed in partnership with Memorial Sloan Kettering Cancer Center in New York City that helps oncologists identify personalized, evidence-based treatment options based on massive volumes of data. This breakthrough technology is now helping scale access to knowledge at Bumrungrad International Hospital in Thailand, Manipal Hospitals in India, and more than 20 hospitals in China. Cognitive assistants are at work helping build more intimate, personalized relationships at the Brazilian bank Banco Bradesco, the insurance company GEICO, and retailer The North Face. Dublin-based Medtronic plc, a global health-care solutions company, is creating a cognitive app for people with diabetes to predict a hypoglycemic event hours in advance. These are just a few examples of organizations that are using cognitive systems today.

It's important to note that we are not talking about the AI we see in movies. This isn't about creating a synthetic brain or an artificial human. Rather, this is about augmenting human intelligence. Indeed, there is nothing in either cognitive science or its application that implies either sentience or autonomy.

Of course, anyone familiar with the history of technology knows that technological breakthroughs often have major effects on work and jobs. Some jobs are eliminated, while others are created. With cognitive systems, we are already beginning to see the emergence of new disciplines—from data curation to system training, as well as new fields of scientific knowledge and new kinds of work—quite possibly more than in any prior technology revolution.

Data can be seen as the world's great new resource. What steam power, electricity, and fossil fuels did for earlier eras, data promises to do for the 21st century—if we can mine, refine, and apply it. Thanks to the new generation of cognitive technologies, we can. Intelligence augmentation—IA as opposed to AI—will change how humans work together, make decisions, and manage organizations.

3

Rise of the Strategy Machines

Thomas H. Davenport

As a society, we are becoming increasingly comfortable with the idea that machines can make decisions and take actions on their own. We already have semi-autonomous vehicles, high-performing manufacturing robots, and automated decision making in insurance underwriting and bank credit. We have machines that can beat humans at virtually any game that can be programmed. Intelligent systems can recommend cancer cures and diabetes treatments. "Robotic process automation" can perform a wide variety of digital tasks.

What we don't have yet, however, are machines for producing strategy. We still believe that humans are uniquely capable of making "big swing" strategic decisions. For example, we wouldn't ask a computer to put together a new mobility strategy for a car company based on such trends as a decreased interest in driving among teens, the rise of ride-on-demand services like Uber and Lyft, and the likelihood of self-driving cars at some point in the future. We assume that the defined capabilities of algorithms are no match for the uncertainties, high-level issues, and problems that strategy often serves up.

We may be ahead of smart machines in our ability to strategize right now, but we shouldn't be complacent about our human dominance. First, it's not as if we humans are really that great at setting strategy. Many M&A deals don't deliver value, new products routinely fail in the marketplace, companies expand unsuccessfully into new regions and countries, and myriad other strategic decisions don't pan out.

Second, although it's unlikely that a single system will be able to handle all strategic decisions, the narrow intelligence that computers display today is already sufficient to handle specific strategic problems. IBM Corp., for example, has begun to use an algorithm rather than just human judgment to evaluate potential acquisition targets. Netflix Inc. uses predictive analytics to help decide what TV programs to produce. Algorithms have long been used to identify specific sites for retail stores, and could probably be used to identify regions for expansion as well. Key strategic tasks are already being performed by smart machines, and they'll take on more over time.

A third piece of evidence that strategy is becoming more autonomous is that major consulting firms are beginning to advocate for the idea. For example, Martin Reeves and Daichi Ueda, both of the Boston Consulting Group, published an article in April 2016 on the *Harvard Business Review* website called "Designing the Machines That Will Design Strategy," in which they discuss the possibility of automating some aspects of strategy. McKinsey & Co. has invested heavily in a series of software capabilities it calls "McKinsey Solutions," many of which depend on analytics and the semi-automated generation of insights. Deloitte has developed a set of internal and client offerings involving semi-automated sensing of an organization's external environment. In short, there is clear movement within the strategy consulting industry toward a greater degree of interest in automated cognitive capabilities.

Assuming that this movement toward autonomous strategy is beginning to take place, what are the implications for human strategists? As Reeves and Ueda point out in their article, cognitive capabilities will need to be combined with human intelligence in what they call an "integrated strategy machine." Just as contemporary autonomous vehicles can take the wheel under certain conditions, we'll see situations in which strategic decision making can be automated. Other situations, however, will require that a human strategist take the wheel and change direction.

Big-picture thinking is one capability at which humans are still better than computers—and will continue to be for some time. Machines are not very good at piecing together a big picture in the first place, or at noticing when the landscape has changed in some fundamental way. Good human strategists do this every day.

In a world of smart, strategic machines, humans need to excel at big-picture thinking in order to decide, for example, when automation is appropriate for a decision; what roles machines and people will play, respectively; and when an automated strategy approach their organization has implemented no longer makes sense.

Executives who see the big picture are able to answer the critical questions that will guide their organizations' future: how their companies make money, what their customers really want, how the economy is changing, and what competitors are up to that is relevant to their company.

These kinds of issues and trends can't be captured in data alone. It's certainly a good and necessary thing for strategists to begin embedding their thinking into cognitive technologies, but they also have to keep their eyes on the broader world. There is a level of sense-making that only a human strategist is capable of—at least for now. It's a skill that will be more prized than ever as we enter an era of truly strategic human–machine partnerships.

4

Predicting a Future Where the Future Is Routinely Predicted

Andrew W. Moore

Workers on the factory floor have suddenly gathered at a point along the production line. Some are scratching their heads. Others are gesticulating wildly. Most stand with their hands in their pockets. Something is wrong, and no one has thought to call management.

In the near future, scenes like this one will be obsolete. Thanks to advances in artificial intelligence (AI), managers will be alerted to workplace anomalies as soon they occur. Unusual behaviors will be identified in real time by cameras and image-processing software that continuously analyze and comprehend scenes across the enterprise.

The hunch-based bets of the past already are giving way to far more reliable data-informed decisions. But AI will take this further. By analyzing new types of data, including real-time video and a range of other inputs, AI systems will be able to provide managers with insights about what is happening in their businesses at any moment in time and, even more significantly, detect early warnings of bigger problems that have yet to materialize.

As a researcher, I learned to appreciate the value of early warnings some years ago, while developing algorithms for analyzing data from hospital emergency rooms and drugstores. We discovered that we could alert public health officials to potential epidemics and even the possibility of biological warfare attacks, giving them time to take countermeasures to slow the spread of disease.

Similar analytic techniques are being deployed to detect early signs of problems in aircraft. The detailed maintenance and flight logs for the US Air Force's aging fleet of F-16 fighter jets are analyzed automatically to identify patterns of equipment failures that may affect only a handful of aircraft at present, but have the potential to become widespread. This has enabled officials to confirm and diagnose problems and take corrective action before the problems spread.

With AI, we can have machines look for millions of worrying patterns in the time it would take a human to consider just one. But that capability includes a terrible dilemma: the multiple hypotheses problem. If you sound an alarm whenever something is anomalous at a 99% confidence level, and you check millions of things an hour, then you will receive hundreds of alarms every minute.

Statisticians and AI researchers are working together to identify situations and conditions that tend to sound false alarms, like a truckload of potassium-rich bananas that can set off a radiation detector meant to identify nuclear materials. By reducing the risk of false alarms, it will be possible to set sensor thresholds even lower, enhancing sensitivity.

The predictive benefits of AI will stretch well beyond equipment and process analysis. For instance, researchers are having great success with algorithms that closely monitor subtle facial movements to assess the emotional and psychological states of individuals. Some of the most interesting applications now are in the mental health sphere, but imagine if the same tools could be deployed on checkout lines in stores, lines at theme parks, or security queues at airports. Are your customers happy or agitated? Executives wouldn't need to wait weeks or even days for a survey to be completed; these systems could tell you the emotional state of your customers right now.

Other researchers are deploying AI in the classroom. When I taught, I couldn't tell whether the lecture I was giving was any good—at least not when it would still benefit me or my students. But simple sensors like microphones and cameras can be used by AI programs to detect when active learning is taking place. Just the sounds alone—Who's talking? Who isn't? Is anyone laughing?—can provide a lot of clues about teaching effectiveness and when adjustments should be made.

Such a tool could also be used to gauge whether your employees are buying in to what you're sharing with them in a meeting, or if potential customers are engaged during focus groups. A "managerial Siri" might take this even further. If you asked your digital assistant, "Do the folks in my staff meeting seem to be more engaged since we had the retreat?" you might receive an answer such as, "Yes, there is an increase in eye contact between team members and a slight but significant increase in laughter."

As a manager, I absolutely detest being surprised. And like everyone else, despite the petabytes of data at my fingertips, I too often am. But AI doesn't get overwhelmed by the size and complexity of information the way we humans do. Thus, its promise to keep managers more in the know about what's really happening across their enterprise is truly profound.

5

Using Artificial Intelligence to Set Information Free

Reid Hoffman

Artificial intelligence (AI) is about to transform management from an art into a combination of art and science. Not because we'll be taking commands from science fiction's robot overlords, but because specialized AI will allow us to apply data science to our human interactions at work in a way that earlier management theorists like Peter Drucker could only imagine.

We've already seen the power of specialized AI in the form of IBM's Watson, which trounced the best human players at "Jeopardy," and Google DeepMind's AlphaGo, which recently defeated one of the world's top Go players, Lee Sedol, four games to one. These specialized forms of AI can process and manipulate enormous quantities of data at a rate our biological brains can't match. Therein lies the applicability to management: Within the next five years, I expect that forward-thinking organizations will be using specialized forms of AI to build a complex and comprehensive corporate "knowledge graph."

Just as a social graph represents the interconnection of relationships in an online social network, the knowledge graph will represent the interconnection of all the data and communications within your company. Specialized AI will be ubiquitous throughout the organization, indexing every document, folder, and file. But AI won't stop there. AI will also be sitting in the middle of the communication stream, collecting all of the work products, from emails to files shared to chat messages. AI will be able to draw the connection between when you save a proposal, share it with a colleague, and discuss it through corporate messaging. This may sound a bit Big Brother-ish, but the result will be to give knowledge workers new and powerful tools for collecting, understanding, and acting on information.

Specialized AI will even help us improve that scourge of productivity, the meeting. Meetings will be recorded, transcribed, and archived in a knowledge repository. Whenever someone in a meeting volunteers to tackle an action item, AI software will record and track those commitments, and automatically connect the ultimate completion of that item back to the original meeting from whence it sprang. Sound far-fetched? The AI techniques for classification, pattern matching, and suggesting potentially related information are already part of our everyday lives. You encounter them every time you start typing a query into Google's search box, and the autocomplete function offers a set of choices, or every time you look at a product on Amazon, and the site recommends other products you might like.

The rise of the knowledge graph will affect the practice of management in three key ways:

Better Organizational Dashboards Right now, organizational dashboards—the sets of information executives monitor and use to guide decision making—are limited to structured data that is easy to extract or export from existing systems, such as revenues, app downloads, and payroll information. These backward-looking metrics do have value: They help managers understand what has happened in their operations and identify hot spots for troubleshooting. But AI-generated knowledge graphs will dramatically expand the scope of these dashboards. For example, managers will be able to access sentiment analysis of internal communications in order to identify what issues are being most discussed, what risks are being considered, and where people are planning to deploy key resources (whether capital or attention). AI-powered dashboards will provide forward-looking, predictive intelligence that will deliver a whole new level of insight to managerial decision making. Computers won't be making decisions for us, but they can sift through vast amounts of data to highlight the most interesting things, at which point managers can drill down, using human intelligence, to reach conclusions and take actions. This is an example of what Joi Ito, director of the MIT Media Lab, refers to as "extended intelligence"—in other words, treating intelligence as a network phenomenon and using AI to enhance, rather than replace, human intelligence.

Data-Driven Performance Management Current performance management processes are terribly flawed. A Deloitte study found that just 8% of organizations believe that their annual review process excels at delivering business value. One of the big reasons for this dissatisfaction is the lack of data to drive objective performance management. In order to manage performance, you have to be able to measure it, and in most organizations, this simply isn't possible for the majority of employees. Senior leaders might be evaluated based on the company's overall performance, and certain functions like sales have objective, quantitative performance metrics, but almost everyone else is evaluated by subjective criteria and analysis. In the absence of data, internal politics and unconscious bias can play a major role, resulting in performance management that is biased and inaccurate. The knowledge graph will allow managers to identify the real contributors who are driving business results. You'll be able to tell who made the key decision to enter a new market and which people actually took care of the key action items to make it happen. Yet even as the knowledge graph reduces the role of guesswork and intuition, the human manager will still be in the loop, exercising informed judgment based on much better data. The result will be much more efficient allocation of human capital, as people are better matched with projects that suit their strengths, and the best people are deployed against the highest-leverage projects.

Increased Talent Mobility As we get better at allocating human capital, organizations will need to do a better job of supporting increased talent mobility, both inside and outside the organization. In the networked age, talent will tend to flow to its highest-leverage use. Each such "tour of duty" will benefit both the company and the worker. But people are not plug-and-play devices; they need time to become productive in a new role (in part because it takes time to build the needed connections into a new network). The knowledge graph will make onboarding and orientation far more rapid and effective. On the very first day on the job, a worker will be able to tap the knowledge graph and understand not just his or her job description, but also the key network nodes he or she will need to work with. Rather than a new employee simply being handed a stack of files, onboarding AI software will be able to answer questions like, "Whom do I need to work with on the new office move? What were the meetings where it was discussed? When is our next status meeting?" The new employee will also be able to ask how things were done in the past (for example, "Show me a tag cloud of the topics my predecessor was spending his time on. How has that allocation evolved over the past 12 months?"). AI might even ask outgoing employees to review and annotate the key documents that should be passed on to their successors. The tacit knowledge that typically takes weeks or months to amass in today's workplace will have been captured in advance so that within the first hours of accepting a new job, an employee will be able to start apply-ing that knowledge.

For all AI's potential benefits, some very smart people are worried about its potential dangers, whether they lie in creating economic displacement or in actual conflict (such as if AI were to be applied to weapons systems). This is precisely why I am, along with friends like Sam Altman, Elon Musk, Peter Thiel, and Jessica Livingston, backing the OpenAI project, to maximize the chances of developing "friendly" AI that will help, rather than harm, humanity. AI is already here to stay. Leveraging special-ized AI to extend human intelligence in areas like management is one way we can continue to progress toward a world in which artificial intelligence enhances the future of humanity.

6

What to Expect from Artificial Intelligence Technology

Ajay Agrawal, Joshua S. Gans, and Avi Goldfarb

Major technology companies such as Apple, Google, and Amazon are prominently featuring artificial intelligence (AI) in their product launches and acquiring AI-based startups. The flurry of interest in AI is triggering a variety of reactions—everything from excitement about how the capabilities will augment human labor to trepidation about how they will eliminate jobs. In our view, the best way to assess the impact of radical technological change is to ask a fundamental question: How does the technology reduce costs? Only then can we really figure out how things might change.

To appreciate how useful this framing can be, let's review the rise of computer technology through the same lens. Moore's law, the long-held view that the number of transistors on an integrated circuit doubles approximately every two years, dominated information technology until just a few years ago. What did the semiconductor revolution reduce the cost of? In a word: *arithmetic*.

This answer may seem surprising since computers have become so widespread. We use them to communicate, play games and music, design buildings, and even produce art. But deep down, computers are souped-up calculators. That they appear to do more is testament to the power of arithmetic. The link between computers and arithmetic was clear in the early days, when computers were primarily used for censuses and various military applications. Before semiconductors, "computers" were humans who were employed to do arithmetic problems. Digital computers made arithmetic inexpensive, which eventually resulted in thousands of new applications for everything from data storage to word processing to photography.

AI presents a similar opportunity: to make something that has been comparatively expensive abundant and cheap. The task that AI makes abundant and inexpensive is *prediction*—in other words, the ability to take information you have and generate information you didn't previously have. In this essay, we will demonstrate how improvement in AI is linked to advances in prediction. We will explore how AI can help us solve problems that were not previously prediction oriented, how the value of some human skills will rise while others fall, and what the implications are for managers. Our speculations are informed by how technological change has affected the cost of previous tasks, allowing us to anticipate how AI may affect what workers and managers do.

Machine Learning and Prediction

The recent advances in AI come under the rubric of what's known as "machine learning," which involves programming computers to learn from example data or past experience. Consider, for example, what it takes to identify objects in a basket of groceries. If we could describe how an apple looks, then we could program a computer to recognize apples based on their color and shape. However, there are other objects that are apple-like in both color and shape. We could continue encoding our knowledge of apples in finer detail, but in the real world, the amount of complexity increases exponentially.

Environments with a high degree of complexity are where machine learning is most useful. In one type of training, the machine is shown a set of pictures with names attached. It is then shown millions of pictures that each contain named objects, only some of which are apples. As a result, the machine notices correlations—for example, apples are often red. Using correlates such as color, shape, texture, and, most important, context, the machine references information from past images of apples to predict whether an unidentified new image it's view-ing contains an apple.

When we talk about prediction, we usually mean anticipating what will happen in the future. For example, machine learning can be used to predict whether a bank customer will default on a loan. But we can also apply it to the present by, for instance, using symptoms to develop a medical diagnosis (in effect, *predict-ing* the presence of a disease). Using data this way is not new. The mathematical ideas behind machine learning are decades old. Many of the algorithms are even older. So what has changed?

Recent advances in computational speed, data storage, data retrieval, sensors, and algorithms have combined to dramati-cally reduce the cost of machine learning-based predictions. And the results can be seen in the speed of image recognition and language translation, which have gone from clunky to nearly perfect. All this progress has resulted in a dramatic decrease in the cost of prediction.

The Value of Prediction

So how will improvements in machine learning impact what happens in the workplace? How will they affect one's ability to complete a task, which might be anything from driving a car to establishing the price for a new product? Once actions are taken, they generate outcomes.

But actions don't occur in a vacuum. Rather, they are shaped by underlying conditions. For example, a driver's decision to turn right or left is influenced by predictions about what other drivers will do and what the best course of action may be in light of those predictions.

Seen in this way, it's useful to distinguish between the cost versus the value of prediction. As we have noted, advances in AI have reduced the *cost* of prediction. Just as important is what has happened to the *value*. Prediction becomes more valuable when data is more widely available and more accessible. The computer revolution has enabled huge increases in both the amount and variety of data. As data availability expands, prediction becomes increasingly possible in a wider variety of tasks.

Autonomous driving offers a good example. The technology required for a car to accelerate, turn, and brake without a driver is decades old. Engineers initially focused on directing the car with what computer scientists call "if then else" algorithms, such as "If an object is in front of the car, then brake." But progress was slow; there were too many possibilities to codify everything. Then, in the early 2000s, several research groups pursued a useful insight: A vehicle could drive autonomously by predicting what a human driver would do in response to a set of inputs (inputs that, in the vehicle's case, could come from camera images, information using the laser-based measurement

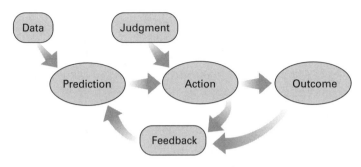

The Anatomy of a Task

Actions are shaped by the underlying conditions and the resolution of uncertainty to lead to outcomes. Drivers, for example, need to observe the immediate environment and make adjustments to minimize the risk of accidents and avoid bottlenecks. In doing so, they use judgment in combination with prediction.

method known as LIDAR, and mapping data). The recognition that autonomous driving was a prediction problem solvable with machine learning meant that autonomous vehicles could start to become a reality in the marketplace years earlier than had been anticipated.

Who Judges?

Judgment is the ability to make considered decisions—to understand the impact different actions will have on outcomes in light of predictions. Tasks where the desired outcome can be easily described and there is limited need for human judgment are generally easier to automate. For other tasks, describing a precise outcome can be more difficult, particularly when the desired outcome resides in the minds of humans and cannot be translated into something a machine can understand.

This is not to say that our understanding of human judgment won't improve and therefore become subject to automation. New modes of machine learning may find ways to examine the relationships between actions and outcomes, and then use the information to improve predictions. We saw an example of this in 2016, when AlphaGo, Google's DeepMind artificial intelligence program, succeeded in beating one of the top players in the world in the game of Go. AlphaGo honed its capability by analyzing thousands of human-to-human Go games and playing against itself millions of times. It then incorporated the feedback on actions and outcomes to develop more accurate predictions and new strategies.

Examples of machine learning are beginning to appear more in everyday contexts. For instance, x.ai, a New York City–based artificial intelligence startup, provides a virtual personal assistant for scheduling appointments over email and managing calendars. To train the virtual assistants, development team members had the virtual assistants study the email interactions between people as they schedule meetings with one another so that the technology could learn to anticipate the human responses and see the choices humans make. Although this training didn't produce a formal catalog of outcomes, the idea is to help virtual assistants mimic human judgment so that over time, the feedback can turn some aspects of judgment into prediction problems.

By breaking down tasks into their constituent components, we can begin to see ways AI will affect the workplace. Although the discussion about AI is usually framed in terms of machines versus humans, we see it more in terms of understanding the level of judgment necessary to pursue actions. In cases where whole decisions can be clearly defined with an algorithm (for example, image recognition and autonomous driving), we expect

In cases where whole decisions can be clearly defined with an algorithm, we expect to see computers replace humans.

to see computers replace humans. This will take longer in areas where judgment can't be easily described, although as the cost of prediction falls, the number of such tasks will decline.

Employing Prediction Machines

Major advances in prediction may facilitate the automation of entire tasks. This will require machines that can both generate reliable predictions and rely on those predictions to determine what to do next. For example, for many business-related language translation tasks, the role of human judgment will become limited as prediction-driven translation improves (though judgment might still be important when translations are part of complex negotiations). However, in other contexts, cheaper and more readily available predictions could lead to increased value for human-led judgment tasks. For instance, Google's Inbox by Gmail can process incoming email messages and propose several short responses, but it asks the human judge which automated response is the most appropriate. Selecting from a list of choices is faster than typing a reply, enabling the user to respond to more emails in less time.

Medicine is an area where AI will likely play a larger role—but humans will still have an important role, too. Although artificial intelligence can improve diagnosis, which is likely to lead to more effective treatments and better patient care, treatment and care will still rely on human judgment. Different patients have different needs, which humans are better able to respond to than machines. There are many situations where machines may never be able to weigh the relevant pros and cons of doing things one way as opposed to another way in a manner that is acceptable to humans.

The Managerial Challenge

As artificial intelligence technology improves, predictions by machines will increasingly take the place of predictions by humans. As this scenario unfolds, what roles will humans play that emphasize their strengths in judgment while recognizing their limitations in prediction? Preparing for such a future requires considering three interrelated insights:

Prediction is not the same as automation. Prediction is an input in automation, but successful automation requires a variety of other activities. Tasks are made up of data, prediction, judgment, and action. Machine learning involves just one component: prediction. Automation also requires that machines be involved with data collection, judgment, and action. For example, autonomous driving involves vision (data); scenarios—given sensory inputs, what action would a human take? (prediction); assessment of consequences (judgment); and acceleration, braking, and steering (action). Medical care can involve information about the patient's condition (data); diagnostics (prediction); treatment choices (judgment); bedside manner (judgment and action); and physical intervention (action). Prediction is the aspect of automation in which the technology is currently improving especially rapidly, although sensor technology (data) and robotics (action) are also advancing quickly.

The most valuable workforce skills involve judgment. In many work activities, prediction has been the bottleneck to automation. In some activities, such as driving, this bottleneck has meant that human workers have remained involved in prediction tasks. Going forward, such human involvement is all but certain to diminish. Instead, employers will want workers to augment the value of prediction; the future's most valuable skills will be those that are complementary to prediction—in other words, those related to judgment. Consider this analogy: The demand for golf balls rises if the price of golf clubs falls, because golf clubs and golf balls are what economists call complementary goods. Similarly, judgment skills are complementary to prediction and will be in greater demand if the price of prediction falls due to advances in AI. For now, we can only speculate on which aspects of judgment are apt to be most vital: ethical judgment, emotional intelligence, artistic taste, the ability to define tasks well, or some other forms of judgment. However, it seems likely that organizations will have continuing demand for people who can make responsible decisions (requiring ethical judgment), engage customers and employees (requiring emotional intelligence), and identify new opportunities (requiring creativity).

Judgment-related skills will be increasingly valuable in a variety of settings. For example, if prediction leads to cheaper, faster, and earlier diagnosis of diseases, nursing skills related to physical intervention and emotional comfort may become more important. Similarly, as AI becomes better at predicting shopping behavior, skilled human greeters at stores may help differentiate retailers from their competitors. And as AI becomes better at anticipating crimes, private security guards who combine ethical judgment with policing skills may be in greater demand. The part of a task that requires human judgment may change over time, as AI learns to predict human judgment in a particular context. Thus, the judgment aspect of a task will be a moving target, requiring humans to adapt to new situations where judgment is required.

The future's most valuable skills will be those that are complementary to prediction – in other words, those related to judgment.

Managing may require a new set of talents and expertise. Today, many managerial tasks are predictive. Hiring and promoting decisions, for example, are predicated on prediction: Which job applicant is most likely to succeed in a particular role? As machines become better at prediction, managers' prediction skills will become less valuable while their judgment skills (which include the ability to mentor, provide emotional support, and maintain ethical standards) become more valuable.

Increasingly, the role of the manager will involve determining how best to apply artificial intelligence, by asking questions such as: What are the opportunities for prediction? What should be predicted? How should the AI agent learn in order to improve predictions over time? Managing in this context will require judgment both in identifying and applying the most useful predictions, and in being able to weigh the relative costs of different types of errors. Sometimes there will be well-acknowledged objectives (for example, identifying people from their faces). Other times, the objective will be less clear and therefore require judgment to specify the desired outcome. In such cases, managers' judgment will become a particularly valuable complement to prediction technology.

Looking Ahead

At the dawn of the 21st century, the most common prediction problems in business were classic statistical questions such as inventory management and demand forecasting. However, over the last 10 years, researchers have learned that image recognition, driving, and translation may also be framed as prediction problems. As the range of tasks that are recast as prediction problems continues to grow, we believe the scope of new applications will be extraordinary. The key challenges for executives will be (1) shifting the training of employees from a focus on prediction-related skills to judgment-related ones; (2) assessing the rate and direction of the adoption of AI technologies in order to properly time the shifting of workforce training (not too early, yet not too late); and (3) developing management processes that build the most effective teams of judgment-focused humans and prediction-focused AI agents.

The authors wish to thank James Bergstra, Tim Bresnahan, and Graham Taylor for helpful discussions. All views remain our own.

7

The Shifts – Great and Small – in Workplace Automation

David H. Autor

There have been periodic warnings in the last two centuries that automation and new technology would wipe out large numbers of middle-class jobs.

In the early 19th century, for instance, a group of English textile artisans, known as Luddites, famously protested the automation of textile production by seeking to destroy some of the machines. A century later, concern rose again over "The Automation Jobless," as they were called in the title of a *Time* magazine story of Feb. 24, 1961. US President Lyndon B. Johnson even empaneled a Commission on Technology, Automation, and Economic Progress to confront the productivity problem in 1964—specifically, that productivity was rising so fast it might outstrip demand for labor. The Commission ultimately concluded that automation did not threaten employment, but that didn't permanently close the case.

Employment displacement concerns are valid and have regained prominence. For instance, in their widely discussed book, *The Second Machine Age*, MIT scholars Erik Brynjolfsson and Andrew McAfee offer an unsettling picture of the likely effects of automation on employment.

While we can say with certainty that the past two centuries of automation and technological progress have not made human labor obsolete—the employment-to-population ratio actually rose during the 20th century as women moved from home to market—past interactions between automation and employment do not necessarily predict the future. There's no fundamental economic law that guarantees every adult a living solely on the basis of sound mind and good character.

In particular, the emergence of greatly improved computing power, artificial intelligence (AI), and robotics raises the possibility of replacing labor on a scale not previously observed. If this should occur, the primary challenge will be one of income distribution rather than nonemployment. How can we ensure that the largest number of people gain from the surge in productivity?

Labor Market Polarization

Automation does, indeed, substitute for labor—as it is typically intended to do. However, automation also *complements* labor, raises output in ways that lead to higher demand for labor, and interacts with adjustments in labor supply.

The frontier of automation is rapidly advancing, yet the challenges to fully substituting machines for workers in tasks requiring flexibility, judgment, and common sense remain immense. In many cases, machines both substitute for and complement human labor—substituting for workers in routine, codifiable tasks, while amplifying the comparative advantage of workers in problem-solving skills, adaptability, and creativity. Focusing only on what is lost misses a central economic mechanism by which automation affects the demand for labor: It raises the value of the tasks that workers uniquely supply.

The biggest challenge as new technologies emerge is in the types of jobs created and what those jobs pay—not the number of jobs per se. Although automation may not prove the enemy of employment, it may pose a large challenge for income distribution. In the last few decades, a "polarization" of the labor market is emerging in which employment gains accrue disproportionately in jobs at the top and bottom of the distribution. The rapid employment growth in both high- and low-education jobs has substantially reduced the share of employment for "middle-skill" jobs. In 1979, the four middle-skill occupations— sales, office and administrative workers, production workers, and operatives—accounted for 60% of employment. In 2007, this number was 49%, and in 2012, it was 46%.

What specific job changes can we expect going forward? Consider the surprising complementarities between information technology (IT) and employment in banking. When automated teller machines (ATMs) were introduced, their numbers in the US economy quadrupled from approximately 100,000 to 400,000 between 1995 and 2010. You might assume that these machines all but eliminated bank tellers in that interval. But US bank teller employment actually rose modestly from 500,000 to approximately 550,000 over the 30-year period from 1980 to 2010.

IT enabled a broader range of bank personnel to become involved in "relationship banking" as the routine cash-handling tasks of bank tellers receded. Banks recognized that tellers could be both checkout clerks and salespeople, forging relationships with customers and introducing them to additional bank services like credit cards, loans, and investment products. This is an example of a fundamental economic reality that is frequently overlooked: *Tasks that cannot be substituted by automation are generally complemented by it.*

Job Quantity versus Quality

Even if automation does not reduce the quantity of jobs, it may greatly affect the quality of available jobs. As the price of computing power has fallen, computers and their robot cousins have increasingly displaced workers in accomplishing explicit, codifiable tasks.

Tasks that have proved most vexing to automate are those demanding flexibility, judgment, and common sense—skills that we understand only tacitly. In particular, two broad sets of tasks have proven stubbornly challenging to computerize.

One category requires problem-solving capabilities, intuition, creativity, and persuasion. These tasks, termed "abstract," are characteristic of professional, technical, and managerial occupations. They employ workers with high levels of education and analytical capability, and they place a premium on inductive reasoning, communications ability, and expert mastery.

The second category includes tasks requiring adaptability, visual and language recognition, and in-person interactions— which we call "manual" tasks. Manual tasks are characteristic of food preparation and serving jobs, cleaning and janitorial work, grounds cleaning and maintenance, in-person health assistance, and numerous jobs in security and protective services.

Workers in abstract, task-intensive occupations benefit from IT because of strong complementarities between routine and abstract tasks, elastic demand for services, and inelastic labor supply to these occupations over the short and medium term— even though these activities present daunting challenges for automation. IT, then, should raise earnings in occupations that make intensive use of abstract tasks and among workers who intensively supply them.

The Role of Machine Learning

When considering how human labor can complement new technology, I see two distinct paths: environmental control and machine learning. The first regularizes the environment, so that comparatively inflexible machines can function semi-autonomously. In the second approach, engineers develop machines that attempt to infer rules from context, abundant data, and applied statistics.

Some researchers expect that as computing power rises and training databases grow, the machine-learning approach will meet or exceed human capabilities. Others suspect that machine learning will only "get it right" on average, while missing many of the most important and informative exceptions.

My prediction is that employment polarization will not continue indefinitely. While some of the tasks in many middle-skill jobs are susceptible to automation, many will continue to demand a mixture of tasks from across the skill spectrum. For example, medical support occupations—radiology technicians, phlebotomists, nurse technicians, and others—are a significant and rapidly growing category of relatively well-remunerated, middle-skill employment. A significant stratum of middle-skill jobs, combining vocational skills with literacy, numeracy, adaptability, problem solving, and common sense, will persist in coming decades.

This prediction has one obvious catch: the ability of the US education and job-training system (both public and private) to produce the kinds of workers who will thrive in these jobs of the future. In this and other ways, the issue is not that middle-level workers are doomed by automation and technology, but instead that human-capital investment must be at the heart of any long-term strategy for producing skills that are complemented, rather than substituted, by technological change.

8

How Blockchain Will Change Organizations

Don Tapscott and Alex Tapscott

For the last century, academics and business leaders have been shaping the practice of modern management. The main theories, tenets, and behaviors have enabled managers to build corporations, which have largely been hierarchical, insular, and vertically integrated. However, we believe that the technology underlying digital currencies such as bitcoin—technology commonly known as blockchain—will have profound effects on the nature of companies: how they are funded and managed, how they create value, and how they perform basic functions such as marketing, accounting, and incentivizing people. In some cases, software will eliminate the need for many management functions.

Sound far-fetched? Let us explain. The Internet vastly improved the flow of data within and between organizations, but the effect on how we do business has been more limited. That's because the Internet was designed to move information—not value—from person to person. When you email a document, photograph, or audio file, for example, you aren't sending the original—you're sending a copy. Anyone can copy and change it. In many cases, it's legal and advantageous to share copies.

By contrast, if you want to expedite a business transaction, emailing money directly to someone is not an option—not only because copying money is illegal but also because you can't be 100% certain the recipient is the person he says he is. As a result, we use intermediaries to establish trust and maintain integrity. Banks, governments, and in some cases big technology companies have the ability to confirm identities so that we can transfer assets; the intermediaries settle transactions and keep records.

For the most part, intermediaries do an adequate job, with some notable exceptions. One concern is that they use servers that are vulnerable to crashes, fraud, and hacks. Another is that they often charge fees—for example, to wire money overseas. They also monitor customer behavior and collect data, and they exclude the hundreds of millions of people who can't qualify for a bank account. And sometimes, they make terrible mistakes, as the 2008 financial crisis made evident.

What would happen if there were an Internet of value where parties to a transaction could store and exchange value without the need for traditional intermediaries? In a nutshell, that's what blockchain technology offers. Value isn't saved in a file somewhere; it's represented by transactions recorded in a global spreadsheet or ledger, which leverages the resources of a large peer-to-peer network to verify and approve transactions. A blockchain has several advantages. First, it is distributed: It runs on computers provided by volunteers around the world, so there is no central database to hack. Second, it is public: Anyone can view it at any time because it resides on the network. And third, it is encrypted: It uses heavy-duty encryption to maintain security.

Blockchain transactions are continuously verified, cleared, and stored by the network in digital blocks that are connected to preceding blocks, thereby creating a chain. Each block must refer to the preceding block to be valid. This structure permanently time-stamps and stores exchanges of value, preventing anyone from altering the ledger. To steal anything of value, a thief would have to rewrite its entire history on the blockchain. Collective self-interest ensures the blockchain's safety and reliability. Therefore, we think blockchain provides a powerful mechanism for blowing traditional and centralized models, such as that of the corporation, to bits.

The Role of Transaction Costs

In a classic article published in 1937 titled "The Nature of the Firm," economist Ronald H. Coase noted that there are costs associated with organizing production through the open market rather than through a firm—such as the cost of searching for relevant prices and the cost of negotiating numerous contracts. Coase expected businesses to expand internally until the cost of performing an additional transaction inside the organization become equal to the cost of using the open market. In a 1976 article, scholars Michael C. Jensen and William H. Meckling added another dimension by introducing the concept of "agency costs," which are the costs associated with managers' tendencies to make decisions that are not optimal from an owner's point of view.

We believe that blockchain will transform how businesses are organized and managed. It allows companies to eliminate transaction costs and use resources on the outside as easily as resources on the inside.

Like many other analysts, we envisioned that the Internet would reduce transaction costs so that corporate boundaries would become more porous and organizations would seek talent outside their boundaries. As it turned out, the costs fell much further than we expected and in turn lowered barriers to entry for startups and established businesses looking to expand into adjacent areas. To be sure, the Internet reduced the costs of search, while email, social media, cloud computing, and applications such as enterprise resource planning reduced the costs of coordination. More broadly, these new capabilities enabled corporations to outsource overhead, crowdsource innovation, and eliminate middle managers and other intermediaries, thus freeing industries such as accounting, commercial banking, and even music to consolidate assets and operations.

Managing With Blockchain

We believe that blockchain will transform how businesses are organized and managed. It allows companies to eliminate transaction costs and use resources on the outside as easily as resources on the inside. Vertical integration may continue to make sense in some situations (for manufacturing controlled pharmaceuticals, for example, or where companies have industry-leading strengths throughout the supply chain). But in most cases, we believe that networks based on blockchain will be better suited for creating products and services and for delivering value to stakeholders.

Human Resources and Procurement

Blockchain will enable organizations requiring specialized talent and capabilities to obtain better information about potential contractors and partners than many traditional recruitment and procurement methods offer. With a prospective employee's consent, an employer will have access to a cache of information that's known to be correct because it has been uploaded, stored, and managed on a highly secure, distributable database. For example, job prospects wouldn't be able to lie about their training or degrees because an authority, such as the university they graduated from, has entered the data on the blockchain. Tampering with data after the fact wouldn't be possible: It would involve taking over the entire blockchain, a nearly impossible task. Individuals would control their own personal data (including birth date, citizenship, financials, and educational records) in a virtual black box. They alone would be able to decide what to do with the information.

Human resources and procurement staff will need to learn how to query the blockchain with specific yes or no questions—for example, Do you have this kind of license? Can you code in this specific language? The responses from all the black boxes will provide a list of people who meet these qualifications. Employers can ask whatever they want, and job seekers can program their black boxes with answers or refuse to answer.

Finance and Accounting

Information about a business's financial well-being changes all the time. When you search the web for a company's financial data, you search in two dimensions: horizontal (across the web) and vertical (within particular websites). What you find can be out-of-date or inaccurate in other ways. On a blockchain, though, there's a third dimension: sequence. In addition to being able to obtain a historical picture of the company since it was incorporated, you can see what has occurred in the last few minutes. The opportunity to search a company's complete record of value will have profound implications for transparency as it brings to light off-book transactions and hidden accounts. People responsible for records and reports will be able to create filters that allow stakeholders to find what they are searching for at the press of a button. Companies will be able to create transaction ticker tapes and dashboards, some for internal use and some for the public. As extreme as this may sound, it's really not.

Sales and Marketing

Just as a blockchain provides a way to obtain information about potential contractors and partners, it will be able to tell you about people or businesses who are potential customers. As we have noted, individuals will control access to their own data in virtual black boxes, which will limit a company's ability to profile customers by tracking and capturing their behavior online. However, the blockchain will allow companies to engage with individual customers on a peer-to-peer basis.

This may seem like a lot of effort, but it could actually be a huge opportunity. Some consumers may offer businesses access to their data in exchange for freebies; others will charge fees to license their data. Either way, companies will be able to reach their target audience with greater precision.

What's more, sellers won't have to worry about who the customers are and whether they are able to pay. With the new platform, sellers won't have to incur the cost of establishing trust—thus they can facilitate transactions that would have been risky or might not have been possible otherwise. Furthermore, blockchains will eliminate the cost of warehousing data and protecting other people's data from security breaches. It should also be easier to target customers who make their interests known.

Despite the advantages of being able to reduce risk, there is also a downside. The ability to make precise queries leads to

We believe that block-chain will transform how businesses are organized and managed. It allows companies to eliminate transaction costs and use resources on the outside as easily as resources on the inside.

With the new platform, sellers won't have to incur the cost of establishing trust – thus they can facilitate transactions that would have been risky or might not have been possible otherwise.

precise results. This means that there will be much less serendipity. With blockchains, you are less likely to discover people or partners who don't fit your profile but are open to change, willing to adapt, and eager to learn.

Legal Affairs

Coase and subsequent economic theorists have argued that corporations are vehicles for creating long-term contracts when short-term contracts require too much effort to negotiate and enforce. Blockchains facilitate contracting in both the short and long term. Through smart contracts—software that, in effect, mimics the logic of contracts with guaranteed execution, enforcement, and payments—companies will be able to automate the terms of agreement. A contract can refer to data fields elsewhere on the blockchain (for example, a party's account balance, a change in a commodity price, or an additional sale of a copyrighted work). It can trigger alerts and ensure payments.

Because the contracts will be self-enforcing, corporations will not want to enter into them lightly. Changing the terms of deals (or attempting to manipulate them) will be more challenging. Lawyers and other managers will need to learn how to audit legal templates and make sure the contract software supports what both parties agreed to do. They will also need to become knowledgeable on issues involving the blockchain and smart contracts. The fastest-growing specialty in the law firm of the future is likely to be "smart contract mediator."

Raising Capital

We believe blockchains will also transform the process of raising money. In our view, the blockchain has the potential to disrupt the way the global financial system works and change the nature of investment. Mindful of this prospect, the New York Stock Exchange has invested in Coinbase Inc., a digital currency wallet and platform company headquartered in San Francisco, California. For its part, the Nasdaq Stock Market is also experimenting with blockchain technology.

Shareholders will be able to enforce the commitments executives make. Companies can specify relationships and state specific outcomes and goals so that everyone understands what the respective parties have signed up to do.

Integrating the Pieces

So how will blockchain help companies become stronger competitors? How can a company use it to integrate the various pieces? Blockchain technology provides a platform for people to work together with the persistence and stability of an organization but without the hierarchy. Consider ConsenSys Inc., a venture production studio based in Brooklyn, New York, that builds decentralized software applications and end-user tools that operate on blockchain. Founder Joseph Lubin describes the company's structure as a hub-and-spoke arrangement rather than hierarchical; each project operates on its own, with the major contributors holding equity. For the most part, people get to choose what they work on. The central hub provides supporting services to the spokes in exchange for a share of the ownership. The various rights and relationships are codified in smart contracts that hold the entity together.

In recent years, we have been reminded all too often that managers don't always act with the highest degree of integrity. (Think of the scandals at Enron, AIG, and Volkswagen, for instance.) What if we could codify ethics and integrity into the circuitry

Shareholders will be able to enforce the commitments executives make. Companies can specify relationships and state specific outcomes and goals so that everyone understands what the respective parties have signed up to do.

of the enterprise, or reduce the moral hazard that too often sees management gambling with shareholder capital? Through smart contracts under blockchain, shareholders will be able to enforce the commitments executives make. Companies can specify relationships and state specific outcomes and goals so that everyone understands what the respective parties have signed up to do and whether those things are actually getting done.

On blockchain, executives will someday no longer need to attest that their books are in order once a year or every quarter; the blockchain will keep a company's books in order in what is, in effect, real time as a matter of course. Financial statements will go from snapshots of the enterprise at one point in time to a transparent, three-dimensional view of the whole enterprise. Shareholders and regulatory agencies alike will be able to examine the books whenever they choose. Institutional investors will have the ability to create their own credit dashboards based on the facts, as opposed to relying on interpretations by ratings agencies. And ratings agencies themselves may overhaul their rating systems based on information from blockchains.

In contrast to the Internet, which took two decades to develop and yet another decade to become commercial, the blockchain ecosystem is developing more rapidly as an economic platform. For executives, this means there is little time to waste. They will want to examine their industries and their competitors with an eye toward identifying opportunities for profitable growth.

Executives should begin by identifying people within the company who are interested in the technology or using digital currency. They should talk to people in the company's IT department about the technology's implications, buy some bitcoin, and experiment with purchasing inexpensive items on the blockchain to see how it works. At the same time, they should identify nearby companies using blockchain—take the opportunity to visit their operations and talk with people involved, and invite experts to meet with the team. Now is the time to reimagine how your company organizes the way it creates value. If you don't, someone else will.

9

Is Your Company Ready to Operate as a Market?

Rita Gunther McGrath

New technologies are eroding transaction costs—and in the process creating a world that is increasingly connected. The resulting level of interdependence creates a radically new set of challenges for management.

In particular, complicated business situations are being replaced with complex ones. In a complicated system, even though there may be many inputs and outputs, one can predict the outcome by knowing how the system works. For instance, the global air travel system is complicated, and yet its unprecedented safety has been made possible by driving down the margin of error and correcting known defects. In a complex system, on the other hand, different parts can interact in ways that make predicting, and therefore controlling, the outcome nearly impossible. For instance, the interconnectedness of the global financial system means that events occurring in one part of the system have unexpected interactions with others, leading to unpredictable outcomes. As Janet L. Yellen, now chair of the US Federal Reserve System, said in 2013, "Complex links among financial market participants and institutions are a hallmark of the modern global financial system. Across geographic and market boundaries, agents within the financial system engage in a diverse array of transactions and relationships that connect them to other participants." Such interconnectedness was blamed for both the severity of the 2008 financial crisis and the difficulties encountered in resolving it.

Connecting parts of a system that used to be sealed off from one another can create enormous benefits. For instance, companies installing enterprise resource management (ERM) systems benefit from having different operations across silos able to share information. Companies using various electronic payment systems benefit from decreased costs of doing business. In a complex system, however, benefits for one set of players can create losses for others.

Consider, for instance, what shipping was like before the 1950s-era invention of the shipping container. It required tens of thousands of dockworkers to load and unload ships. Not only was this a time-consuming process, but it also left cargo vulnerable to theft, as it was very difficult to track the contents of an individual load. The introduction of a container that could be sealed at the factory, shipped, and then transported inland by train or truck transformed global trade. *The Economist* in 2013 reported that "new research suggests that the container has been more of a driver of globalization than all trade agreements in the past 50 years taken together." According to *The Economist*, one study of industrialized countries found that containerization was associated with "a 320% rise in bilateral trade over the first five years after adoption and 790% over 20 years. By comparison, a bilateral free-trade agreement raises trade by 45% over 20 years and GATT [General Agreement on Tariffs and Trade] membership adds 285%."

The dramatic fall in the cost of shipping fundamentally altered the assumptions management had accepted as given until that time. Once it became possible to ship even low-value goods and make a profit, the rules of competition changed. Rather than being a relatively fixed commodity located in one physical place—the docks—labor could now be sourced from anywhere containers could be packed. And rather than employer and worker being tied to each other in one relatively enduring relationship, an army of outsourced and freelance workers came into play and redefined the dynamics between management and labor. The advent of shipping containers created global competition for jobs and transformed entire supply chains. As former Intel chairman and CEO Andy Grove lamented in 2010, the unintended consequences of all this were to undermine job creation in the United States, even as employment growth among US trade partners in Asia skyrocketed.

The markets vs. hierarchies concept, originally pioneered by the economist Oliver E. Williamson, suggested the conditions under which one could operate purely by contracting on an open market as opposed to requiring some kind of organization (a hierarchy) to accomplish one's goals. Hierarchies are favored, in his formulation, when uncertainty is high, various parties face a risk of opportunism in market exchanges, and information about what is being exchanged is asymmetrically distributed.

As with containerization 60 years ago, new ways of sharing information today and, most likely, robotics and data analytics in the near future have the effect of continuing to push transactions that were once executed within an organization's boundaries out into open markets. Consider how clearly the Ubers and Airbnbs of the world today demonstrate that one does not need to own assets in order to use them. And this phenomenon goes well beyond those popular examples, as the rise of Amazon Web Services and the proliferation of software-as-a-service offers demonstrates. The advent of blockchain technology further promises to distribute tasks that used to be centralized within organizations into markets—by offering an alternative to the centralized validation banks provide, for example.

How does management attention need to shift as the world moves more toward market forms of organizing? Clearly, we are moving from a business world dominated by hierarchies, in which assets are controlled by a company, to a world of markets, in which assets can be accessed when needed. The conventional relationship between buyers and suppliers then shifts to more complex configurations in multisided markets and ecosystems. Networks become a primary vehicle for exchanging information of all kinds. Increasingly porous organizational boundaries mean that information is less likely to be hoarded. And increasingly, customers are looking to organizations for complete experiences rather than product and service features.

Managing in such a complex environment requires not only traditional management skills such as planning and controlling, but also new ones, such as negotiating complex agreements, quickly detecting the unexpected, accelerating organizational learning, and fostering the creation of trusting relationships among groups and teams who may be only temporarily associated with one another. And this all takes place at an accelerated pace of change for which many will be unprepared. The lines between a defined managerial role in a traditional company and an entrepreneurial role in these newly emerging market contexts are definitely blurring.

Practically, what does this mean for managers looking for a new way of operating? First, the assumption that change is the unusual thing and stability is the normal thing is worth challenging. Today, leaders need to identify, as early as possible, the patterns that deserve their attention and make constant course-correcting adjustments. Instead of being precise but slow and reinforcing existing perspectives, leaders need to be comfortable with making roughly right and fast decisions and with challenging the status quo. Instead of just using traditional management tools such as net present value, managers need to be more discovery-driven and options-oriented. And they need to remember that one of the most valuable gifts colleagues can give one another in a complex environment is candor about what is really going on out there.

10

The End of Corporate Culture as We Know It

Paul Michelman

We are evolving toward an age of networked enterprises, in which the traditional hierarchies of the corporation will be supplanted by self-organizing systems collaborating on digital platforms.

It will be the era of entrepreneurship, distributed leadership, and the continual reorganization of people and resources. It will be the time of disintermediation both within and between organizations. Layers of management will fall; the need for centralized systems and trusted go-betweens will dissipate, if not disappear.

Or so many experts predict.

As for me? Yes, I do believe this is the future toward which we are slowly advancing. Those of us deeper into our careers may not see it come to full fruition during our organizational lives, but the trends are real, and they are already on display if you care to look for them.

And that makes me genuinely worried for my friends in the corporate-culture business. Because I'm not sure that culture is going to matter all that much in the future—at least not in the ways we conceive of it today.

Jon Katzenbach, one of the field's most respected and creative thinkers, defines an organization's culture as "the self-sustaining pattern of behavior that determines how things are done." It is "made of instinctive, repetitive habits and emotional responses." Culture is meant to provide a well-rooted sense of purpose within an organization, exemplified by a recognized set of behaviors and shared beliefs. It gets—and keeps—everyone marching in the same direction.

Creating and maintaining culture is, thus, painstaking work. It demands focus and commitment throughout organizations. During my work life, I have been lifted by strong corporate cultures and nearly drowned by weak ones. I have no doubt of culture's power to align an organization and enliven its workforce. And there are plenty of studies to back that up.

But that's history speaking.

As to life in the digital matrix, there is reason to question culture's role. Our relationships to institutions will become increasingly defined by the activity in which we are engaged at any given time. We will come to view ourselves as "affiliates" more than "employees" as we think of that term today. We will encounter new partners and colleagues on a rolling basis. We will weave in and out of relationships and flow across broad platforms of commerce in individual and small-group nodes, working interchangeably with people who belong to the same organization and those who do not.

In this world, we will no longer prize alignment; we will prize realignment.

Such an environment benefits from clear and universal rules of engagement. It does not benefit from habits that are distinctive to one group of people—which is the essence of organizational culture.

In his 1966 book, *The Will to Manage*, the celebrated father of management consulting Marvin Bower described a company's philosophy as "the way we do things around here." Those words helped to establish the role of corporate culture and to solidify its purpose over the next 50 years. But we are embarking upon a time when the "way we do things" will be reinvented with each new collaboration on the network. And in these waters, a tool meant to reinforce consistency of behavior over long periods of time transforms from a motor to an anchor.

11

Do You Have a Conversational Interface?

Bala Iyer, Andrew Burgert, and Gerald C. Kane

We live in the age of mobile applications. There are currently several million apps available. This profusion of choices means it can be difficult for users not only to choose which apps to download, but to manage them all—a phenomenon we call "app fatigue." This situation creates both a need and an opportunity to engage users on a single platform. Today, that platform is increasingly becoming messaging apps.

We think the next era will belong to "the conversational layer"—both text- and voice-driven—that will use chat, messaging, or natural language interfaces to interact with people, brands, services, and bots. This shift is currently evidenced by the massive adoption of messaging apps such as Facebook Messenger, Echo, and WhatsApp, which together host more than 60 billion messages daily. According to eMarketer, messaging apps will reach 2 billion people within a few years. WhatsApp users average nearly 200 minutes each week using the service, and many teenagers now spend more time on smartphones sending instant messages than perusing social networks.

Messaging platforms can also alter the way businesses can communicate with their customers. Currently, conversational interfaces within well-known messaging platforms such as Facebook Messenger, Slack, Skype, WeChat, Kik, and Telegram allow companies to chat with their users.

Bots as Conversation Partners

While mobile chat platforms are interesting, the arrival of artificial intelligence (AI)–powered engines called bots have made them a powerful tool for sense-making and commerce. Bots use machine-learning techniques to understand text and provide better responses to user queries. They are present in the background, and they make sense of the conversations taking place and convert them into actions using apps, such as scheduling a meeting or ordering a pizza. For example, imagine you are chatting with your business partner using Messenger and discussing a visit to a client site in Boston. Using machine-learning algorithms, a bot can recognize that you are talking about travel and initiate a transaction with your favorite travel app, such as Expedia, or offer a link for a ride through Uber. The messaging platform effectively becomes a distribution channel for software and services without leaving the conversation.

1-800-Flowers recently launched such an experience on Facebook Messenger and has since expanded to Amazon Alexa and the IBM Watson platforms. Customers can order flowers directly from their experience in these conversational layers. 1-800-Flowers is very focused on customer support and maintaining a relationship with their customers, so the company jumped at the opportunity to be one of the first in the space. Of the tens of thousands of people who have ordered flowers through the chatbot integration, more than 70% are new customers—and these new customers skew toward younger demographics than the company's existing customers.

What Should Companies Expect?

Companies should position themselves for the conversational layer to be more widespread five to 10 years from now. Individual users will most likely want to interact with trusted brands to fulfill their needs through natural language interactions. This interaction will occur at the exact time the user demands a product or service, and in the exact terms she thinks of that product or service, in the language and communication methods she typically uses (intent, words, shortcuts, emojis, etc.). Companies need to strengthen these natural language capabilities in their products, apps, and bots to allow users to communicate with them with ease.

Individuals will begin to welcome and even expect this type of service from brands, but companies must remember that this trusted personal space is precious. Poorly designed interactions can irreparably damage the customer relationship. For example, when Microsoft's Tay posted racist remarks on Twitter, it had to be shut down temporarily. The bot industry as a whole has yet to come up with the "killer bot" that tips the scale for wider adoption, but bots continue to grow in sophistication and power.

What Should Companies Do Today?

Pick a platform When reaching users, a brand needs to understand where current or potential customers spend their chat time. The platform choice is an important early decision. This is similar to how brands and engineering teams initially opted to launch their products on iOS, then Android, and other mobile OS platforms early on. In order to reach user conversations today, brands will need to decide which platforms to target and build on. Different platforms have a diverse set of capabilities (i.e., user identity, cards, and buttons on Messenger, and work team and slash commands on Slack) and demographics target.

Run strategic experiments It is not clear if customers would use the conversational layer for quick responses or for broader conversations. Brands like Amex Finance are using chatbots to provide notifications to customers about their new products or alerts about forthcoming travel dates. These limited experiments would allow Amex to set the bar for the nature of interaction with clients. The enterprise social networking platform Slack is using chatbots to automate routine managerial check-ins, reducing the need for meetings.

Look for innovative uses in other sectors Simple examples like ordering airline tickets or pizzas are emerging. However, sophisticated bots that understand the context and make intelligent decisions have not yet been developed. In fact, recent articles on automated email assistants have shown that they rely too much on human intervention.

Companies can also learn from examples outside their industry. For example, a Georgia Tech professor used a bot as a teaching assistant for a programming class. Using machine-learning techniques, the bot was able to handle 97% of the student queries. In online education, where dropout rates are quite high, a high level of engagement with a "tutor" could make a huge difference in retention.

Pilot bots with your customers   Many of the tools that are provided by the messaging and bot platform providers are from the open-source space, and companies can perform low-cost experiments with a reduced set of users to learn more about conversational interactions and use cases that yield the desired results.

The conversational layer of computing may have not yet fully arrived, but it is coming. Companies should begin thinking and experimenting now about how to use these new avenues to support their brand and their business today, so they can be ready for that conversational future as users demand and engage with this type of experience.

12

Unleashing Creativity with Digital Technology

Robert D. Austin

If you watch movies or television, you've likely seen Stefan Sonnenfeld's work. It's on display in *Star Wars: Episode VII–The Force Awakens*, three *Mission Impossible* movies, four *Transformer* movies, the *Cold Case* TV series, and dozens more. In a 2007 article, *Entertainment Weekly* listed him alongside creative luminaries like Steven Spielberg and Meryl Streep as one of the "50 Smartest People in Hollywood." Unlike some others on the list, however, Sonnenfeld's creativity is digitally enabled.

Sonnenfeld is a "digital intermediate colorist." He uses computers to alter the colors in movies and TV shows until they look spectacular. Without the technology, his artistry would not be possible. Nor would the profits that Sonnenfeld and the company he cofounded, Company 3, have generated for its parent companies, Ascent Media (from 2000 to 2010) and Los Angeles–based Deluxe Entertainment Services Group Inc. (to which it was sold in 2010).

Digital color artistry like Sonnenfeld's is an example of a general principle: Technology can be deployed to augment the creative abilities of people and organizations and make new and valuable forms of innovation possible. Today's digital technologies have reached a level of maturation that enables, across many domains, a practical capability that I have, in my research, called *cheap and rapid iteration*.

To iterate is to try something different from what you tried last time. Sonnenfeld iterates when he tries out many different color effects on a movie. Sometimes he tries this, then that, then another thing, until he hits on something brilliant. He can do this cheaply and rapidly only because he works at a high-powered computer console with fiber connections, huge amounts of storage, and specialized software for making subtle adjustments to specific areas of a picture and across time in a film. Yes, all that equipment is an investment, but once he's made it, the cost of trying something new—of the next iteration—is nominal.

Iteration is the process that enables most forms of artistry. Painters often create numerous versions of a painting; Pablo Picasso, for example, created dozens of "studies" prior to his famous *Guernica* painting. Theater artists rehearse, trying a scene this way, then that. Designers iterate by building quick-and-dirty prototypes.

Processes often become more creative when rapid iteration is affordable. Unfortunately, this is not the case in a lot of business domains; often, in business, it's costly to try something new, especially if it doesn't work out. That was the case in the film-coloring business before digital technologies, when the process involved painting directly onto film or all-or-nothing photochemical processes. Just as digital changed the game in film coloring, so it can in many areas of business where it's been too expensive to experiment much.

It's most natural to think about such opportunities in product businesses, using technologies like computer-aided design (CAD) and 3-D printing to try out a design and then tweak it.

But the phenomenon is also occurring in service businesses. A bank I know is using social media technologies to do quick and rapid trials of new customer offerings. To pull this off, a company must use digital technologies to lower the cost of the processes associated with the iterative cycle: setting up a new try (reconfiguration), trying it (simulation or testing), and examining and interpreting the results (visualization).

In the next five years, managers will awaken to a wide range of new possibilities. They'll act to improve creative capabilities, by figuring out how to deploy technologies to replace expensive physical trying with cheap virtual trying. In effect, they'll be constructing virtual rehearsal spaces, virtual laboratories, and inexpensive prototyping facilities. The aim won't be to design machines to take over people's jobs, but rather to augment human capabilities.

This is not a new idea. In the 1960s, Doug Engelbart proposed using computers to augment human intellect. Also in the 1960s, Internet pioneers J. C. R. Licklider and Robert W. Taylor emphasized the potential of computer networks to enhance creative work. Michael Schrage, in the late 1990s, described how simulation allowed companies to engage in "serious play." My own recent research and that of others point to a coming new age of organizational creativity—an era that may, I believe, finally be here.

National Public Radio once called Sonnenfeld the "da Vinci of the movies." In the coming few years, managers will begin to realize that they can create such da Vincis throughout their companies. And that will be a very profitable thing to do.

13

Rethinking the Manager's Role

Lynda Gratton

I've been thinking about technology and management for over a decade. In the process, I have written two books describing some of the ways the practice of management will respond to rapid technological innovations. Looking back, I made four predictions about management and technology.

First, it was clear to me that the manager's role as a coordinator of work would come under increasing pressure. Constant improvements in robotics and machine learning, in conjunction with the automation of routine tasks, make management a more unclear practice. What is a manager, and what is it that managers do? Are we witnessing the end of management?

Next, I could see an inevitable shift in which a parent-to-child way of looking at the relationship between the manager and his or her team would be questioned and ultimately superseded by an adult-to-adult form. The nexus of this more adult relationship concerns how commitments are made and how information is shared. When technology enables many people to have more information about themselves and others, it's easier to take a clear and more mature view of the workplace. Self-assessment tools, particularly those that enable people to diagnose what they do and how they do it, can help employees pinpoint their own productivity issues. They have less need for the watchful eyes of a manager.

Third, it seemed to me obvious that technology would tip the axis of power from the vertical to the horizontal. Why learn from a manager when peer-to-peer feedback and learning can create stronger lateral forms of coaching? Moreover, technology-enabled social networking is capable of creating robust and realistic maps of influence and power—so no more hiding behind fancy job titles.

Finally, the rise of platform-based businesses such as Uber Technologies Inc. has everyone excited about platforms and how they can create a fertile arena for new businesses to be built while also acting as a conduit for flexible ways of working.

What is the role of management in all of this?

From those four predictions, one could easily imagine that "the end of management" is in sight—crushed by peer feedback, pushed out by specialist roles, disintermediated by powerful platforms, and exposed by social network analysis.

And yet it seems that the current reality is a great deal more complex. Rather than seeing the end of management, we seem to be witnessing the rise of a more skilled form of it. Over the last seven years, I have directed the Future of Work Consortium, a group that has involved executives from 60 multinational companies from all over the world and from different sectors. Through workshops, focus groups, and an annual survey, we have followed the impact that technology is having on work and management.

For most of our participants, the surprise is that so little has changed over the last five years in the way they work. In fact, they report that their use of technology in their personal and home life has far exceeded their experience of technology at work. Many indicate that the real positive impact of technology has been on the way they run their everyday life rather than on their productivity at work. Inevitably, this will change in the next five years—but how will management use new technology tools?

We asked these executives to consider how they see the future and to assess the current capability of their corporations. From this, we were able to identify the "future risk factors"—those aspects of the corporation that will be important in the future but are currently poorly executed. The same top risks came out every year: how to manage virtual teams; how to manage multigenerational groups (particularly with regard to differences in technology use); and how to support rapid knowledge flows across business units.

Notice what all three areas of risk have in common: They are all fundamentally about management. But this is a very complex form of management—managing virtually rather than face to face; managing when the group is diverse rather than homogenous; and managing when the crucial knowledge flows are across groups rather than within. These are highly skilled roles in terms of both managerial capabilities (for example, how to build rapid trust, coach, empathize, and inspire) and management practices (for example, team formation, objective setting, and conflict resolution). It is these managerial skills and practices that will be augmented by technology over the coming years in ways we may not yet have grasped—but that will emerge over time.

14

The Three New Skills Managers Need

Monideepa Tarafdar

In the coming years, both business leaders and their employees will face a number of challenges as they deal with changing digital technologies. In particular, they will need to learn three important new skills: how to partner with new digital "colleagues"; how to create a mindful relationship with increasingly ubiquitous digital technologies; and how to develop empathy for the varying technology preferences of their human coworkers. Organizations, for their part, will need to design programs and processes to support these efforts.

Partner with Digital "Colleagues"

Employees across a wide spectrum of industries will be working with what are, in effect, "digital coworkers"—algorithms that help them tackle a range of tasks such as answering call-center help desk questions, making financial investment decisions, diagnosing medical conditions, scheduling and running manufacturing assembly lines, and providing dashboard advice regarding important performance indicators. These digital colleagues will embody intelligence that evolves cognitively and learns continuously about the specific task it is applied to, by incorporating new solutions learned from experience and applying them to future problems.

Given the complexity and often real-time application of this sort of intelligence, it may be unnecessary and indeed impossible for human professionals to verify the veracity of an algorithm's solutions. However, as the data become denser and algorithms get faster and more complex, there is a danger of "runaway algorithms" that become disconnected from the reality of the phenomenon they represent, eventually leading to wrong solutions. To prevent this, managers will need to retain their expertise and control over their tasks and processes. They should provide context for the decisions and recommendations of their digital partners by monitoring those decisions from time to time and recalibrating them against their own experience, insight, and intuition—even going against their digital coworkers if necessary.

While digital colleagues will, for the most part, independently handle routine aspects of their tasks, exceptions—that is, those cases where their digital intelligence does not have a satisfactory solution—will require human decision making. At the same time, cloud-based intelligent algorithms for relatively narrow and contained tasks—for example, understanding niche buyer behavior—will make it possible for managers to solve everyday problems more effectively.

Leveraging such opportunities will require managers to be alert to opportunities and problems, to have deep process knowledge and to explore, innovate, and engage with their digital coworkers. In short, managers will be confronted and challenged by digital colleagues—just as they are by their human coworkers. They will need to learn how and when to question, agree, compromise, and stretch.

Become Digitally Mindful

Because digital technologies enable remote work, the 9-to-5 workday is becoming less and less meaningful in many settings. Ironically, current management mindsets still focus on the separation of work and nonwork time. Consequently, because managers find it difficult to establish boundaries between work and nonwork, organizations face the fallouts of "technostress," technology addiction, and information overload. However, technologies will increase in flexibility, richness, and seamlessness, and that will lead to their greater use at home for work and vice versa.

The emphasis on work-home conflict ignores the possibilities of such flexibility. It points employees toward managing a conflict rather than leveraging work-home seamlessness. Technology use that enables a continuous flow of meaningful tasks—irrespective of whether they are work-related or not—may be more beneficial for managers' well-being and productivity. Managers should start thinking about cultivating a mindful relationship with the technology—one that embodies their individual preferences about what constitutes such flow. Rather than being troubled about work-home boundaries, which perhaps cannot be maintained in the future, organizations will need to support employees in managing the possibilities of flexibility. The paradigm should shift from conflict to flexibility, from technology detox to flow-driven use, and from the digital dark side to digital mindfulness.

Develop Empathy for Others' Technology Preferences

Even as leaders and managers learn how to work with digital colleagues, they will need to understand and develop empathy for the technology choices and preferences of their human coworkers. A colleague recently objected to me writing work emails to her late at night. On seeing the time stamp the next morning, she felt pressured to answer my messages immediately, to the exclusion of other, more important emails. My first reaction—that she was supposed to prioritize her own email and not be perturbed by what I did with mine—is typical of current organizational mindsets about technology use. Individual managers are so busy managing their own use of technology that they have given little, if any, thought to the preferences and habits of coworkers. However, this goes against an important tenet of management, which is that individuals work best together in teams and departments when there is some level of fit along important aspects.

Everyone has different preferences and habits for using technology. Some may prefer to be contacted by text, others by email, still others by phone or face-to-face. Some may prefer the flexibility afforded by constant email connectivity, while others may favor allotted email time. A clash between preferences can break down communication between teammates and increase misunderstanding, conflict, and stress. Going forward, managers need to not only be proactive about communicating their own technology preferences but also be empathetic about their coworkers' choices, particularly when they are working on the same teams and projects.

In terms of future work design, employees with similar preferences should ideally be put on the same projects and teams. For instance, individuals who like multitasking might appreciate frequent synchronous interactions on instant messaging systems when working on a team together. Those who enjoy constant connectivity might work well with supervisors who share such preferences. More generally, the key to handling this and other similar workplace challenges brought about by digital technologies is for managers to be both flexible and thoughtful in the way they respond.

15

A New Era of Corporate Conversation

Catherine J. Turco

If you want to see how management is changing, take a look inside today's high-tech offices. In the past, corporate leaders sat behind closed doors in large private suites. Today, many sit side by side with employees in open workspaces. In the past, workers toiled alone in cubicles, waiting for formal meetings to speak with their managers and colleagues. Today, they turn and chat with the managers and colleagues sitting right next to them, while conversing with others on digital chat systems that connect the entire organization, and with yet others in lounge areas and cafés built to promote informal connection and dialogue.

These changes are surface manifestations of a deeper transformation underway: Long-held assumptions about corporate communication and hierarchy are breaking down. Social media tools allow more open communication up, down, and across the corporate hierarchy. In the coming years, the savviest leaders will tap into the spirit and tools of openness from social media to build what I call *conversational firms*.

Over the past decade, social media has transformed how people communicate in their personal lives. It is beginning to do the same in our work lives. Millennials who grew up on Twitter,

Facebook, Instagram, Snapchat, and the like are now the fastest-growing portion of the labor force. They are accustomed to constant connection and information access and engage in more open sharing than generations past. And they are carrying these expectations and habits into the workplace.

Meanwhile, the last several years have seen an explosion of social media tools designed for use inside companies—everything from wikis and microblogs, to multichannel platforms such as Yammer, Slack, and HipChat, to employee feedback tools such as TinyPulse. With workers who increasingly expect to have their voices heard, and with tools to enable that, it is now possible—perhaps even paramount—to build more conversational firms.

Conversational firms differ from conventional bureaucratic ones by having a far more open communication environment. Executives use multiple platforms to share information with the entire workforce. They encourage employees to speak up, ask questions, and share ideas and opinions. They saturate the workplace with digital tools and physical spaces designed to encourage dialogue. The result is an ongoing conversation that transcends the formal hierarchical structure.

Forward-thinking leaders are already managing their organizations this way. I profile one such company in my book, *The Conversational Firm: Rethinking Bureaucracy in the Age of Social Media*. When this company's SaaS (software as a service) business model was jeopardized by a spike in customer churn, executives used an internal wiki to share 138 pages of detailed analysis with the full 600+ person workforce. The analysis included the sort of information other executives might pore over in closed meetings but hesitate to share with more than a few select employees—such as trends in bookings, customer acquisition costs, competitors' churn rates, fine-grained profit and loss and cash flow statements, and the complete results of a recent customer survey.

Armed with this information, employees throughout the business responded on the wiki with questions, thoughts, and analyses. The company then extended the conversation offline, hosting an open "Hack Night" for interested individuals to present their ideas in person. Executives sat in the audience listening to employee suggestions and joined small breakout groups to discuss various proposals. In the following weeks, these breakout groups continued to hack away at the churn problem, coordinating their work over the company's internal chat system and posting updates on the wiki for everyone to see. Within a few months, the churn problem was resolved and the organization had new and improved internal processes to avoid future spikes.

This was not an isolated event. During the 10 months I was embedded inside the company, executives shared information broadly and encouraged employees to offer their input on a range of topics. This fostered a well-informed and engaged staff, willing and able to share their knowledge and insights. In turn, the organization was able to respond rapidly and thoughtfully as problems and opportunities arose.

What organization wouldn't want this? In today's markets, customer preferences evolve quickly, and new technologies and competitors continually emerge to unsettle the status quo. Open dialogue is one of the few ways to surface a multidimensional understanding of complex new realities and possible responses. Conversation brings the entire organization's collective wisdom to bear on issues, and in doing so it helps the organization adapt and learn.

Some managers will worry that giving employees broad freedom of speech will weaken their decision-making authority. But blasting open the communication hierarchy doesn't have to mean destabilizing all of the corporate hierarchy. Leaders can give employees voice and engage them in dialogue while retaining the right to make the final call. For example, the company I studied maintained a conventional reporting and decision-making hierarchy while supporting the sort of radically open communication I described above. In fact, the decisions executives made had more legitimacy with the workforce because employees had been invited into the conversation and knew their voices had helped shape the decision-making context. In the customer churn case, executives retained final authority over which ideas the company pursued. However, the ideas themselves were better because they had been shaped by the group's insights and opinions. When done right, open communication can complement formal control.

Doing it right is hard, though. There are challenges to creating conversational companies. For one thing, executives need to see that the point of conversation is to surface a range of opinions. This awareness must guide their approach to everything from the people they hire to how they lead. After all, for valuable dialogue to occur, an organization can't hire a workforce of clones who think and act exactly alike; a diversity of perspectives is required. What's more, employees need to know that their leaders won't punish them for expressing dissenting opinions; absent that trust, people will say only what they think management wants to hear.

Through it all, executives and managers will need patience and a thick skin. Some comments will sting, not every thought shared will be an insightful one, and some conversations will get derailed; these are unavoidable costs. However, leaders willing to invest in truly open dialogue with their workforce will be well positioned to face—and shape—the ever-evolving future.

16

Ethics and the Algorithm

Bidhan L. Parmar and R. Edward Freeman

Are we designing algorithms, or are algorithms designing us? How sure are you that you are directing your own behavior? Or are your actions a product of a context that has been carefully shaped by data, analysis, and code?

Advances in information technology certainly create benefits for how we live. We can access more customized services and recommendations; we can outsource mundane tasks like driving, vacuuming floors, buying groceries, and picking up food. But there are potential costs as well. Concerns over the future of jobs have led to discussions about a universal basic income—in other words, a salary just for being human. Concerns over the changing nature of social interaction have covered topics ranging from how to put your phone down and have a face-to-face conversation with someone to the power dynamics of a society where many people are plugged into virtual reality headsets. Underlying these issues is a concern for our own agency: How will we shape our futures? What kind of world will information technology help us create?

Advances in information technology have made the use of data—principally data about our own behaviors—ubiquitous in the online experience. Companies tailor their offerings based on the technology we employ—for example, the travel website Orbitz a few years ago was discovered to be steering Mac users to higher-priced travel services than it was PC users. Dating sites like eHarmony and Tinder suggest partners based on both our stated and implied preferences. News stories are suggested based on our previous reading habits and our social network activities. Yahoo, Facebook, and Google tailor the order, display, and ease of choices to influence us to spend more time on their platforms, so they can collect even more data and further intermediate our daily transactions.

Increasingly, our physical world is also being influenced by data. Consider self-driving cars or virtual assistants like Siri and Amazon's Echo. There are even children's toys like Hello Barbie that listen, record, and analyze your child's speech and then customize interactions to fit your child.

As our lives become deeply influenced by algorithms, we should ask: What kind of effect will this have?

First, it's important to note that the software code used to make judgments about us based on our preferences for shoes or how we get to work is written by human beings, who are making choices about what that data means and how it should shape our behavior. That code is not value neutral—it contains many judgments about who we are, who we should become, and how we should live. Should we have access to many choices, or should we be subtly influenced to buy from a particular online vendor?

Think of the ethical challenges of coding the algorithm for a self-driving car. Under certain unfortunate circumstances, where an accident cannot be avoided, the algorithm that runs the car will presumably have to make a choice about whether to sacrifice its occupants or risk harming—maybe even fatally—passengers in other cars or pedestrians. How should developers write this code? Despite our advances in information technology, data collection, and analysis, our judgments about morality and ethics are just as important as ever—maybe even more important.

We need to figure out how to have better conversations about the role of purpose, ethics, and values in this technological world, rather than simply assuming that these issues have been solved or that they don't exist because "it's just an algorithm." Questions about the judgments implicit in machine-driven decisions are more important than ever if we are to choose how to live a good life. Understanding how ethics affect the algorithms and how these algorithms affect our ethics is one of the biggest challenges of our times.

17

Why Digital Transformation Needs a Heart

George Westerman

Three technology-driven forces are transforming the nature of management. Automation is making it more and more possible for companies to do work without humans involved. Data-driven management supplements intuition and experience with data and experimentation. Resource fluidity matches tasks to the people who can best perform them, whether inside or outside the organization.

Taken together, these three forces are helping leaders rethink the way work is organized and managed. Tasks that were previously considered the sole domain of humans—like handling customer requests, driving vehicles, or writing newspaper articles—can now be done by machines. Employees at all levels have the information they need to make decisions and adjust their practices. Computers can diagnose situations and identify challenges that humans don't see. Real-time information makes it possible to run experiments rather than guessing what might work. Employees can self-organize, obtain help from experts inside and outside the organization to get a job done. And companies can manage fluctuations in their resource needs through outsourcing, whether through long-term relationships, hourly hiring, or gigs and piecework.

On the whole, these forces are a good thing. They will help managers increase productivity, innovation, and customer satisfaction in the coming years. However, if you lead a traditional company, be careful not to let these forces push your management approach to extremes. Taken to their logical conclusion, the three digital forces could transform management for the worse. Accelerating resource fluidity could make all workers contractors, paid only when the company needs them and earning a living only by cobbling together many different gigs. Data-driven management could become Big Brother, evaluating employees' every action, and hiring or firing people rapidly based only on the numbers. Automation could replace workers and constantly ratchet up the pressure on those that remain. If left unchecked, the three digital forces could transform the employment relationship into an emotionless market transaction—a logically interesting approach that could have negative long-term implications for both workers and companies.

This new employment vision is already starting to play out at some companies. Amazon.com Inc. has an intense, data-driven approach to managing people. As *The New York Times* reported in 2015, it hires only the best, pays them well, works them very hard, and regularly culls its workforce to remove those perceived as lower performing. Uber Technologies Inc. has a relatively small number of very talented, full-time employees, and it engages most of its drivers through contracts that adjust to meet minute-by-minute changes in market demand. Uber is now piloting a fleet of self-driving vehicles.

When making sense of fast-moving digital innovation, it can be tempting to see great born-digital companies like these as aspirational archetypes for management. Certainly, we can—and should—learn a lot from these companies. But think twice before adopting every Silicon Valley management technique as your own. Most companies don't have the resources to attract and pay the world's most talented workers. And many high-performing workers would not be happy in transaction-based work schemes, preferring more security or better work-life balance instead. Traditional companies, even those in nontechnology industries or less-than-sexy locations, can attract great people with the right combination of income, mission, and working conditions.

Beyond the simple question of finding employees, a question remains whether market-based employment contracts are the right kind of social contract for the typical business. These practices, which work well in some fast-growing digital companies, may not be as effective when growth slows or a disruption strikes. Paying people only for time spent on a task reduces opportunities to foster innovation and employee cohesiveness. What's more, such practices do very little to promote loyalty. Many Uber drivers work for its competitors too; they are Uber drivers only until a better offer comes along. Amazon is one of the industry's most innovative companies, but it also has a reputation for high employee turnover. Loyalty helps companies thrive when they cannot pay world-leading wages; employees work hard and create innovations because they believe in the company and its leaders. And when times get tough, loyalty is what helps companies keep their best people.

As a strong proponent of digital transformation, I do not want to discourage traditional companies from adopting digitally powered management practices. However, when building a vision for the future of your company, think of the digital forces as you would vitamins or prescription drugs: The right amounts, applied under the right conditions, can yield fabulous results. But using too much, or in the wrong conditions, can be poisonous.

To summarize, digital transformation needs a heart. In an age of digital innovation, leaders in every industry should strive to transform every part of the company, from customer experience to business models to operational management. But we cannot forget that it is people who make companies work. The vision of management in five or 10 years should not be one where all employees are seen as contracted resources laboring under tight, machinelike supervision. It shouldn't be a world in which automation squeezes workers—and managers—out of the system. It should be one where computers help employees to collaborate fluidly, make decisions scientifically, and manage better with automation than they ever could without it. In the long run, digitally savvy companies that engage the hearts and minds of employees will outperform those that treat people like machines.

18

The Jobs That Artificial Intelligence Will Create

H. James Wilson, Paul R. Daugherty, and Nicola Morini-Bianzino

The threat that automation will eliminate a broad swath of jobs across the world economy is now well established. As artificial intelligence (AI) systems become ever more sophisticated, another wave of job displacement will almost certainly occur.

It can be a distressing picture.

But here's what we've been overlooking: Many new jobs will also be created—jobs that look nothing like those that exist today.

In Accenture PLC's global study of more than 1,000 large companies already using or testing AI and machine-learning systems, we identified the emergence of entire categories of new, uniquely human jobs. These roles are not replacing old ones. They are novel, requiring skills and training that have no precedents.

More specifically, our research reveals three new categories of AI-driven business and technology jobs. We label them trainers, explainers, and sustainers. Humans in these roles will complement the tasks performed by cognitive technology, ensuring that the work of machines is both effective and responsible— that it is fair, transparent, and auditable.

Trainers

This first category of new jobs will need human workers to teach AI systems how they should perform—and it is emerging rapidly. At one end of the spectrum, trainers help natural-language processors and language translators make fewer errors. At the other end, they teach AI algorithms how to mimic human behaviors.

Customer service chatbots, for example, need to be trained to detect the complexities and subtleties of human communication. Yahoo Inc. is trying to teach its language processing system that people do not always literally mean what they say. Thus far, Yahoo engineers have developed an algorithm that can detect sarcasm on social media and websites with an accuracy of at least 80%.

Consider, then, the job of "empathy trainer"—individuals who will teach AI systems to show compassion. The New York-based startup Kemoko Inc., d/b/a Koko, which sprung from the MIT Media Lab, has developed a machine-learning system that can help digital assistants such as Apple's Siri and Amazon's Alexa address people's questions with sympathy and depth. Humans are now training the Koko algorithm to respond more empathetically to people who, for example, are frustrated that their luggage has been lost, that a product they've bought is defective, or that their cable service keeps going on the blink even after repeated attempts to fix it. The goal is for the system to be able to talk people through a problem or difficult situation using the appropriate amount of understanding, compassion, and maybe even humor. Whenever Koko responds inappropriately, a human trainer helps correct that action—and over time, the machine-learning algorithm gets better at determining the best response.

Without the benefit of empathy training Alexa might respond to a user's anxieties with canned, repetitive responses such as, "I'm sorry to hear that" or "Sometimes talking to a friend can help." With the right training, Alexa becomes much more helpful. The following is a verbatim transcription of how Alexa with Koko responds to a person who says he's worried that he'll fail an upcoming exam: "Exams are really stressful, but a little anxiety can help us succeed. It sharpens our minds. ... It's your body's way to prepare itself for action. It's actually a really good thing. I wonder if you could think of your nerves as your secret weapon. Easier said than done, I know. But I think you will do much better than you think."

Explainers

The second category of new jobs—explainers—will bridge the gap between technologists and business leaders. Explainers will help provide clarity, which is becoming all the more important as AI systems' opaqueness increases. Many executives are uneasy with the "black box" nature of sophisticated machine-learning algorithms, especially when the systems they power recommend actions that go against the grain of conventional wisdom. Indeed, governments have already been considering regulations in this area. For example, the European Union's new General Data Protection Regulation, which is slated to take effect in 2018, will effectively create a "right to explanation," allowing consumers to question and fight any decision made purely on an algorithmic basis that affects them.

Companies that deploy advanced AI systems will need a cadre of employees who can explain the inner workings of complex algorithms to nontechnical professionals. For example, algorithm forensics analysts would be responsible for holding any algorithm accountable for its results. When a system makes a mistake or when its decisions lead to unintended negative consequences, the forensics analyst would be expected to conduct an "autopsy" on the event to understand the causes of that behavior, allowing it to be corrected. Certain types of algorithms, like decision trees, are relatively straightforward to explain. Others, like machine-learning bots, are more complicated. Nevertheless, the forensics analyst needs to have the proper training and skills to perform detailed autopsies and explain their results.

Here, techniques like the Local Interpretable Model-Agnostic Explanations (LIME), which explains the underlying rationale and trustworthiness of a machine prediction, can be extremely useful. LIME doesn't care about the actual AI algorithms used. In fact, it doesn't need to know anything about the inner workings. To perform an autopsy of any result, it makes slight changes to the input variables and observes how they alter that decision. With that information, the algorithm forensics analyst can pinpoint the data that led to a particular result.

So, for instance, if an expert recruiting system has identified the best candidate for a research and development job, the analyst using LIME could identify the variables that led to that conclusion (such as education and deep expertise in a particular, narrow field) as well as the evidence against it (such as inexperience in working on collaborative teams). Using such techniques, the forensics analyst can explain why someone was hired or passed over for promotion. In other situations, the analyst can help demystify why an AI-driven manufacturing process was halted or why a marketing campaign targeted only a subset of consumers.

Sustainers

The final category of new jobs our research identified—sustainers—will help ensure that AI systems are operating as designed and that unintended consequences are addressed with the appropriate urgency. In our survey, we found that less than one-third of companies have a high degree of confidence in the fairness and auditability of their AI systems, and less than half have similar confidence in the safety of those systems. Clearly, those statistics indicate fundamental issues that need to be resolved for the continued usage of AI technologies, and that's where sustainers will play a crucial role.

One of the most important functions will be the ethics compliance manager. Individuals in this role will act as a kind of watchdog and ombudsman for upholding norms of human values and morals—intervening if, for example, an AI system for credit approval was discriminating against people in certain professions or specific geographic areas. Other biases might be subtler—for example, a search algorithm that responds with images of only white women when someone queries "loving grandmother." The ethics compliance manager could work with an algorithm forensics analyst to uncover the underlying reasons for such results and then implement the appropriate fixes.

In the future, AI may become more self-governing. Mark O. Riedl and Brent Harrison, researchers at the School of Interactive Computing at Georgia Institute of Technology, have developed an AI prototype named Quixote, which can learn about ethics by reading simple stories. According to Riedl and Harrison, the system is able to "reverse engineer" human values through stories about how humans interact with one another. Quixote has learned, for instance, why stealing is not a good idea and that striving for efficiency is fine except when it conflicts with other important considerations. But even given such innovations, human ethics compliance managers will play a critical role in monitoring and helping to ensure the proper operation of advanced systems.

The types of jobs we describe here are unprecedented and will be required at scale across industries. (For additional examples, see "Representative Roles Created by AI.") This shift will put a huge amount of pressure on organizations' training and development operations. It may also lead us to question many assumptions we have made about traditional educational requirements for professional roles.

Empathy trainers, for example, may not need a college degree. Individuals with a high school education and who are inherently empathetic (a characteristic that's measurable) could be taught the necessary skills in an in-house training program. In fact, the effect of many of these new positions may be the rise of a "no-collar" workforce that slowly replaces traditional blue-collar jobs in manufacturing and other professions. But where and how these workers will be trained remain open questions. In our view, the answers need to begin with an organization's own learning and development operations.

On the other hand, a number of new jobs—ethics compliance manager, for example—are likely to require advanced degrees and highly specialized skill sets. So, just as organizations must address the need to train one part of the workforce for emerging "no-collar" roles, they must reimagine their human resources processes to better attract, train, and retain highly educated professionals whose talents will be in very high demand. As with so many technology transformations, the challenges are often more human than technical.

Representative Roles Created by AI

Trainers

Customer-language tone and meaning trainer	Teaches AI systems to look beyond the literal meaning of a communication by, for example, detecting sarcasm.
Smart-machine interaction modeler	Models machine behavior after employee behavior so that, for example, an AI system can learn from an accountant's actions how to automatically match payments to invoices.
Worldview trainer	Trains AI systems to develop a global perspective so that various cultural perspectives are considered when determining, for example, whether an algorithm is "fair."

Explainers

Context designer	Designs smart decisions based on various factors such as business context, process task, individual, professional, and cultural.
Transparency analyst	Classifies the different types of opacity (and corresponding effects on the business) of the AI algorithms used and maintains an inventory of that information.
AI usefulness strategist	Determines whether to deploy AI (versus traditional rules engines and scripts) for specific applications.

Sustainers

Automation ethicist	Evaluates the noneconomic impact of smart machines, both the upside and downside.
Automation economist	Evaluates the cost of poor machine performance.
Machine relations manager	"Promotes" algorithms that perform well to greater scale in the business and "demotes" algorithms with poor performance.

19

Tackling the World's Challenges with Technology

Andrew S. Winston

The next 5 to 10 years will be a critical time for humanity, in which we can leverage new technologies to address some of our biggest challenges: a dangerously changing climate; pressure on natural resources, especially water and food; and historic levels of inequality.

For decades, business leaders considered these systemic challenges as the domain of government and civil society. Companies, they thought, should focus on creating jobs and making money. But no longer. Expectations are rising fast that business should help tackle our societal challenges—and do it profitably. For example, in one recent study, fully 87% of millennials surveyed believe that the success of business should be measured in more than financial performance.

Technology will clearly shake up how organizations operate in every way—no part of management will go untouched in a world of big data, ubiquitous sensors, and powerful analytics providing new levels of insight. But perhaps the most powerful transformation will change how companies manage their relationships with the larger world.

Business, using technology wisely, will help build a thriving, resilient world. Of course, technology comes with its own environmental and social costs. Better artificial intelligence (AI) and robots could mean fewer jobs. And the "cloud" itself is not actually very light on resources; our IT-based world requires very real—and sometimes hard-to-extract—metals, and it's powered by a lot of energy. But the case for techno-optimism is strong. Consider how new hardware and software will both make businesses run better and improve our lives.

The Internet of things, paired with smarter analytics, will help managers understand a company's impacts on the world in much greater detail. They will know, for example, how much energy and carbon it takes to operate their business at global, brand, geographic, facility, and product levels. Once you measure something, you can manage it. Making better decisions about efficient use of resources is getting easier.

In the largest sectors of the economy, progress is coming quickly. The cost of renewable energy is dropping fast, making clean, distributed energy increasingly competitive with fossil fuels. Physical infrastructure is getting smarter and cleaner as well. New building systems can find heating and cooling systems that someone left on, or shut off lights and computers when nobody is around. These often highly effective operational improvements are becoming trivially easy to achieve.

More powerful computers and cheaper data will also make our transportation systems smarter. Companies with big fleets and good algorithms plot out delivery routes to eliminate unnecessary mileage. Self-driving cars create some operational and moral concerns, but cars in an autonomous system can move much closer together at a steady, energy-saving speed—and they'll save lives. Data and technology are also helping to optimize infrastructure through resource sharing. Car sharing, apartment/room hopping, and worker "hoteling" (which allows an office building to house more people) are saving money and physical resources.

The sector with the largest footprint, food, is undergoing deep change as well. So-called "precision agriculture"—leveraging modern, computerized farm machines—enables farmers to apply fertilizer, pesticide, and water in exact, optimized amounts. And better data on the flow of food through the system should help reduce the monumental food waste that squanders precious embedded energy and water.

But as powerful as these rosy scenarios may seem, I suspect the biggest impact technology will have on societal challenges will be through the radical transparency it enables. New data on supply chains—and generations of workers raised to share everything—will open up everything a company does to public scrutiny. Someday soon, we will find it quaint that Volkswagen AG could conduct a large-scale fraud about their cars' emissions. In the near future, such subterfuge will be impossible. People inside every company will be talking openly about their work on social media, and each car will have sensors collecting real-time emissions data.

Over the coming decade, I believe business will be at the core of helping humanity tackle climate change and manage our shared resources more equitably. Companies will willingly help build a thriving world both to generate profits and because customers, employees, and society will be watching closely and demanding better of them. The way we manage companies is undergoing a deep change: from focusing solely on shareholders to pleasing a broader array of stakeholders, from short-term focus to longer-term strategy, from pursuing operationally narrow goals to embracing collaborative systems thinking.

Technology is both forcing these changes and also offering solutions. That said, we'll also need to change how we use the ultimate technology: our brains. A mindset shift, to rethink the purpose of business and its role in society, will make all the difference.

20

Are You Ready for Robot Colleagues?

Bernd Schmitt, interviewed by Frieda Klotz

Is the convergence between artificial and human intelligence, which once seemed like just a gleam in the eyes of computer scientists and science fiction authors, almost upon us? And if robots become as clever as we are, how will the role of managers change? Bernd Schmitt, the Robert D. Calkins Professor of International Business at Columbia Business School, thinks the convergence is coming, and that managers have to start preparing now.

Schmitt comes at the questions not as a computer scientist but as a marketing expert. He is faculty director of Columbia's Center on Global Brand Leadership, a forum on branding issues for researchers and executives. He conducts research on people's perceptions of cyborgs and robots, and has launched a project entitled "Possible Future Worlds" that explores the impact of technologies on business and consumers. He presents his own brand with a little bit of a lighthearted robotic edge, too: His website is called MeetSCHMITT.com.

Schmitt spoke with *MIT Sloan Management Review* about how artificial intelligence (AI) is advancing, and how it is likely to impact the workplace and even managerial creativity.

MIT Sloan Management Review: For decades, computer scientists have referred to the merging of human intelligence and computing as "the singularity." How do you define the concept?

Bernd Schmitt: Technological singularity is typically defined as a point in the future when IT systems become as sophisticated as humans or even qualitatively more sophisticated and superior to humans. It is often discussed in the context of robots that are supercomputers but also have a human appearance. It's when humans and robots cannot be distinguished from each other because the computer or robot has passed the Turing test, meaning if you ask it a series of questions, it answers like a human being.

It's quite likely that robots, or systems, will at some point have all the capabilities—or more—that we normally ascribe to humans. That includes information processing, of course—in this area computers are already outdoing human decision makers in many respects—but it's not only about decision making. It's also about emotions, it's about creativity, it's about coming up with new ideas.

Now, we are not there yet, obviously. Some futurists, like Ray Kurzweil, predict that this will occur as soon as 20 or 30 years from now. Others argue that it is much further off. But just look at IBM Watson. In 2011, it outperformed humans on the TV game show "Jeopardy!" and today it's doing very, very well with legal documents and with medical diagnoses, and it's outperforming human resource managers. This is a supercomputer—in this case not a robot in human form, but a system—that can do as well as humans or better. And that's the direction of technological singularity.

Robots that have the same emotional and creative capabilities as humans represent the far edge of singularity. Are you one of the people who think we could be seeing that in 20 to 30 years? My view is that when you consider specific tasks like the ones that I just mentioned, providing legal analysis or medical diagnosis and prognoses, coming up with treatment programs and so on—it is certainly possible we will reach that stage in 20 to 30 years, and probably earlier.

Now, if you're talking about technological singularity in the sense that there are systems that are exactly like humans and cannot be distinguished from them, that robots become astute decision makers, become creative and feel emotions, and that their emotional behavior is indistinguishable from that of humans—for that, I think 30 years may be a little bit too optimistic.

But the changes taking place right now are very real. The original robots were just industrial systems. They had an industrial appearance, they didn't look like anything like humans, and they were very mechanical. They were "factory workers." But today we have service robots. There are robots being used in elder care, like Paro, the robot seal, which is used as a psychological aid in Japan. Some robots like Paro are plush, like stuffed animals. Some are moving toward what one could call a human face and a human look.

So, I think in the more emotionally oriented and caring sectors, we will clearly see supercomputing and robots making major inroads.

How would robots be used in the parts of knowledge work that we truly associate with humans, such as decision making, in the modern office?
In the modern office of the near future, as artificial intelligence and robotics incorporate more and more aspects of daily life,

computers will be able to go beyond analytical work and data crunching to carry out more creative, decision-making and emotion-related management tasks.

Human–machine interaction may be quite different from how it is today, with humans in full control. As creative AI systems become more active, robots will make suggestions on how to do work, where to look for answers, how to make decisions, and how to organize and to lead.

Moreover, because the physical form of robots will change and become less mechanical, humans may feel more comfortable around them—although, admittedly, these robots may give some people the creeps. Robots will appear more and more in humanoid forms, and they will display emotions in their faces and move their bodies increasingly naturally.

These robots will have capabilities as idea generators for new product development, as consultants and counsellors, and they may take over many HR roles. They will engage with others in office conversations and office chat. In other words, they may become full-fledged and fully integrated employees and be part of a company's culture. Human office workers will likely end up working next to robot workers.

What impact do you see the singularity having on your areas of specialty – marketing, branding, and creativity?
It's entirely possible that marketing, branding, and creative tasks may be done by supercomputers. Creativity is, basically, knowing a lot about a domain and then making sense of an unexpected event and adjusting to it by devising new solutions. Computers can certainly do that.

How would managerial and leadership roles change as a result of having a partial robot workforce?

Well, if the question is, should managers worry about all of this, the answer is absolutely. I don't think this is about just the front-line employees. Some jobs, like construction work or picking up garbage on the street, may not go away because they involve extremely complicated *motor* behavior. The sophisticated motor tasks that people perform, no computer can do those yet—they're not even close.

But the manager's task is largely *analytical decision making* and, to an extent, dealing with emotions. And much of that is doable by machine. It's actually a very interesting moment in history. We are learning that there are certain skills involved in what people might have seen as simpler jobs that are actually very complicated. And the jobs often viewed as more complex—the knowledge jobs, those involving mental work—actually have a significant automatized component. We see that Watson is as good in certain respects as some doctors and legal analysts, and frankly, I don't think the job of a manager is more complicated than that of a doctor or a lawyer.

So, we should definitely envisage that soon many job functions, and parts of jobs, will be entirely automatized. Some jobs may go away.

In this world of the near future, what can humans do to make themselves useful?

Future generations will live in a world that is much more focused on computing and analytics and big data as well as robots. And as a result, they will have very different relationships with robots. Children will grow up with service robots and having digital or robotic assistants in the house and in professional situations.

As we already know, the idea of a life-long career and moving up a hierarchy of similar jobs has vanished. Humans will need to ask questions such as: How can I structure my career and my job so that I can differentiate myself from a machine? Or, how can I best use my skills to work with an AI system? In other words, humans may supplement the skills of machines—and not the other way around.

What can companies do to prepare?

Company leaders must start developing future scenarios. That's an imperative. It's not a case of waiting five or 10 years to see what happens. This planning should be going on now.

Let me explain why I see this as so urgent. We've seen in just the past 15 years the commercialization of digitization that has entirely changed the workplace and how we communicate. Social media has changed our social lives, and new companies have come up and become million- and billion-dollar businesses in that space—and that has happened in just the last 20 years.

What technologists always stress is that with the speed of technology, the increase is not linear, it's exponential. So, over the next 15 or 20 years, we will see much, much more change, and it will incorporate the digital revolution into solid devices. It's very much this convergence of the digital and the solid that will be happening, and it will be a major life-changing movement and business-changing movement. Businesses need to be ready.

Contributors

Ajay Agrawal is the Peter Munk Professor of Entrepreneurship at the University of Toronto's Rotman School of Management in Toronto, Ontario, as well as founder of the Creative Destruction Lab, a Canada-based training and mentoring program for founders of technology startups.

Robert D. Austin is a professor of information systems at Western University's Ivey Business School in London, Ontario, Canada.

David H. Autor is a Ford Professor of Economics and Associate Department Head at MIT. From 2009 to 2014, he was editor of the *Journal of Economic Perspectives*.

Andrew Burgert is the CEO of Azumo, a San Francisco–based data and AI development firm.

Paul R. Daugherty is Accenture's chief technology and innovation officer.

Thomas H. Davenport is the President's Distinguished Professor of Information Technology and Management at Babson College in Wellesley, Massachusetts, and a fellow of the MIT Sloan Initiative on the Digital Economy.

R. Edward Freeman is University Professor and Olsson Professor of Business Administration at the University of Virginia Darden School of Business.

Joshua S. Gans holds the Jeffrey S. Skoll Chair in Technical Innovation and Entrepreneurship at the Rotman School and is chief economist of the Creative Destruction Lab.

Avi Goldfarb is the Ellison Professor of Marketing at the Rotman School and chief data scientist of the Creative Destruction Lab.

Lynda Gratton is a professor of management practice at London Business School.

Reid Hoffman is executive chairman and cofounder of LinkedIn Corp.

Bala Iyer is a professor and chair of the Technology, Operations, and Information Management Division at Babson College in Wellesley, Massachusetts.

Gerald C. Kane is an associate professor of information systems at the Carroll School of Management at Boston College and the *MIT Sloan Management Review* guest editor for the Digital Business Initiative.

Frieda Klotz is a journalist who writes about technology and health care.

Rita Gunther McGrath, a professor at Columbia Business School, is regarded as one of the world's top experts on strategy and innovation with particular emphasis on developing sound strategy in uncertain and volatile environments.

Paul Michelman is editor in chief of *MIT Sloan Management Review.*

Andrew W. Moore is dean of the School of Computer Science at Carnegie Mellon University in Pittsburgh, Pennsylvania.

Nicola Morini-Bianzino is global lead of artificial intelligence at Accenture.

Tim O'Reilly is the founder and CEO of O'Reilly Media and the organizer of the Next:Economy Summit. He explores the impact of technology on business, work, and society in the weekly Next:Economy Newsletter.

Bidhan L. Parmar is an assistant professor in business administration at the University of Virginia Darden School of Business in Charlottesville, Virginia.

Bernd Schmitt is the Robert D. Calkins Professor of International Business at Columbia Business School. He is also faculty director of Columbia's Center on Global Brand Leadership, a forum on branding issues for researchers and executives.

Alex Tapscott is founder and CEO of Northwest Passage Ventures, an advisory firm incubating early-stage blockchain companies, in Toronto. He is coauthor of *Blockchain Revolution: How the Technology Behind Bitcoin Is Changing Money, Business, and the World* (Portfolio, 2016).

Don Tapscott is the chancellor of Trent University in Peterborough, Ontario, and CEO of the Tapscott Group Inc. in Toronto. He is coauthor of *Blockchain Revolution: How the Technology Behind Bitcoin Is Changing Money, Business, and the World* (Portfolio, 2016).

Monideepa Tarafdar is a professor of information systems at Lancaster University Management School in Lancaster, United Kingdom, and she is currently a visiting scholar at the MIT Sloan Center for Information Systems Research in Cambridge, Massachusetts.

Catherine J. Turco is the Theodore T. Miller Career Development Professor and associate professor of organization studies at the MIT Sloan School of Management, in Cambridge, Massachusetts. She is the author of *The Conversational Firm: Rethinking Bureaucracy in the Age of Social Media* (New York: Columbia University Press, 2016).

Ginni Rometty is chairman, president, and CEO of IBM Corp.

George Westerman is a principal research scientist with the MIT Sloan Initiative on the Digital Economy in Cambridge, Massachusetts.

H. James Wilson is managing director of IT and business research at Accenture Research.

Andrew S. Winston advises many of the world's leading companies on how companies can navigate and profit from the humanity's biggest challenges. He is a globally recognized speaker and writer on business strategy and mega trends. Andrew is the author of *The Big Pivot* and cowrote the international bestseller *Green to Gold*.

Index